McGraw Hill's

CONQUERING

SAT
Math

FOURTH EDITION

Robert Postman

Professor of Mathematics

Ryan Postman

Mathematics
Pascack Hills High School, New Jersey

Elizabeth Postman

Mathematics Teacher and College Instructor

New York Chicago San Francisco Athens London Madrid
Mexico City Milan New Delhi Singapore Sydney Toronto

ISBN   978-1-260-46257-9
MHID   1-260-46257-9

eISBN 978-1-260-46258-6
eMHID   1-260-46258-7

SAT is a registered trademark of the College Entrance Examination Board, which is not involved in the production of, and does not endorse, this product.

Copyright images reproduced with permission from Texas Instruments.

McGraw Hill books are available at special quantity discounts to use as premiums and sales promotions, or for use in corporate training programs. To contact a representative, please visit the Contact Us pages at www.mhprofessional.com.

ACKNOWLEDGMENTS

I am pleased to specially acknowledge the fine efforts of Samantha Lee and Ido Harlev, seniors at NVRHS in Demarest NJ, a top performing public high school. I arranged with publisher support for these exceptionally capable high school students to review a significant portion of the 1000 or so mathematics practice problems in this print/online publication.

Samantha Lee and Ido Harlev were selected because of truly remarkable Mathematics SAT scores and class performance. They worked diligently and cooperatively on a summer project to help you. Their contributions certainly will.

Samantha and Ido support your preparation efforts and wish you every success!

CONTENTS

See the inside front cover of this book for instructions on how to access the bonus online materials.

CHAPTER 1

INTRODUCTION

Conquering SAT Math, Fourth Edition, covers all the mathematics topics and all the problem types on the new version of the SAT, the SAT you will take. Every page in this book was field tested by high school students and reviewed by experienced high school math teachers and test prep experts. We listened very carefully to student and teacher recommendations to create a book that will work for you.

You learn mathematics by doing mathematics. That is why we designed this book and its online resources to have 1000 solved SAT math problems.

We review each of the major areas on the SAT in 26 separate chapters. Each chapter begins with a clear review and practice with answers. Next come several worked-out SAT problems to show you how to apply the mathematics concepts to the SAT format. Then comes a whole set of practice SAT problems in each chapter, with explained answers. The online resources provide over 300 topic based solved SAT problems and over 250 solved problems in SAT test format.

We show you strategies for answering SAT multiple-choice and "grid in" math questions. However, the most important strategy is to think mathematically, just like the people who make up the test do. These test writers always have particular mathematics skills and concepts in mind as they write an item. Knowing how to spot those concepts is usually the secret to getting a high score. You'll learn that from this book.

You Decide How Much Review You Need

This book is designed to help you no matter how much SAT mathematics preparation you need. Which type of student are you?

Many will need all the teaching, the worked out SAT problems, the book's SAT practice problems and the online resources. Some may go right to the book's practice problems and fill in gaps with the online resources. Things will likely change from chapter to chapter. Use the combination of resources you need to meet your goal.

MATHEMATICS TEST OVERVIEW

There are two types of SAT mathematics sections, Math—No Calculator and Math—Calculator. You'll see these two sections one after the other in the second half of the SAT test. Each section contains multiple-choice questions followed by student response (grid-in) questions. Here is an overview of the Math sections of the SAT.

One 25-minute section completed without a calculator

Contains 15 multiple-choice and 5 grid-ins for a total of 20 questions

One 55-minute section completed with a calculator allowed

Contains 30 multiple-choice and 8 grid-ins

1

Math Is Not My Favorite Subject

OK, we understand that. Most of the other parts of the SAT reflect things you might do in a normal week—reading, understanding, and writing. The only time most people do SAT mathematics problems is on the SAT. But that's the way it is. If things ever change, we'll be the first to tell you. Until then, we'll show you the way to your highest possible score.

MATHEMATICS TEST STRATEGIES

There are different strategies for each problem type. Let's take them in order. Before we start here is some overall advice.

Not Sure of an Answer? Skip It and Come Back Later

Skip any question that just seems too difficult. Circle the number of the question in your test booklet and come back later.

MULTIPLE-CHOICE

Remember, a multiple-choice item always shows you the correct answer. Each multiple-choice item has four answer choices. The SAT scores without a guessing penalty, which means that you should guess on every question, even if you cannot eliminate any choices. You have a one in four chance of getting it right anyway!

Process of Elimination

It's sometimes easier to mark out wrong answers until you get to the right one than it is to immediately identify the right answer, so cross out answers as you eliminate them. If you can't get all the way down to one, don't worry! Just guess from the choices remaining after you have eliminated all you can.

Use the Test Booklet

The test booklet is yours. Do your work in the test booklet. Cross off incorrect answers in the test booklet. Draw diagrams—whatever you like.

Darken the Correct Oval

The machine that scores multiple-choice tests does not know if you are right or wrong. The machine just "knows" if you have filled in the correct oval, so it is possible to know the correct answer but be marked wrong.

The fewer times your eyes go from the test sheet to the answer sheet, the less likely you are to make a recording error. We recommend this practice. Write all the answers next to the problem for a two-page spread of questions. Then transfer the answer to the Answer Sheet.

Estimate

Estimate the answer. An estimate may lead you to the correct answer or it may enable you to cross off some answers you know are incorrect.

Work From the Answers

You know that one of the answers must be correct. You may be able to find the correct answer by substituting answers in the problem.

GRID-INS

The answer format for grid-in items is shown below. Write your answer in the four spaces at the top of the grid. But the scoring machine just reads the ovals you fill in. You must fill in the correct oval to receive credit.

You may enter the digits and symbols shown in each column. The symbols represent the fraction line and the decimal point. These choices put some natural limits on the answers. You'll read about those limits in this section.

How to Grid-in an Answer

Here are some examples of grid-in answers. Remember, you must fill in the ovals to receive credit.

Here's how to enter the fraction $\frac{5}{13}$ in the grid. Notice that we entered the numerator and the denominator separated by the slash, and then we filled in the corresponding ovals.

Say that your answer is $\frac{3}{5}$. There are several ways to fill in the grid.

You can grid in $\frac{3}{5}$.   You can grid in the decimal equivalent. 0.6

If your answer is 197, you can fill in the grid two different ways.

The fraction $\frac{2}{3}$ is equivalent to the repeating decimal 0.666 ... You can grid in $\frac{2}{3}$, or .666 or .667.

The Grid Can Help You Determine if Your Answer Is Incorrect

An answer can never be less than 0 or larger than 9999. That means you can never have a negative number as an answer.

The numerator or denominator of a fraction can never have more than two digits.

A decimal can extend only to the thousandths place.

If You Can't Grid in Your Answer, Then Your Answer Is Incorrect

The answer has to fit in the four spaces at the top of the grid. You must be able to select the digits and symbols from the column below each digit or symbol you enter. If you can't do that, your answer is not correct.

THE CALCULATOR

You must bring your own calculator when you take the SAT. A calculator helps you avoid answering questions incorrectly because you make an arithmetic error. However, many students bring calculators that can graph equations. That graphing capability will help with a limited number of SAT questions.

We discuss some calculators below. You should only bring a calculator you know how to use. Don't go out just before the test to buy the latest high-powered calculator; stick with what you know. However, if you know how to use one of the calculators we discuss below, well, so much the better.

Key Entry Errors

We all have a tendency to trust answers that appear on a calculator. However, if you press the wrong keys you'll get an incorrect answer. If the answer is part of a chain of calculations, you probably won't catch it. There are two ways to avoid key entry errors.

Use a Calculator with At Least a Two-Line Display

Most scientific and graphing calculators have at least two lines in the display. That means your entries appear on one line and your answer appears on the other line.

Estimate Before You Calculate

Estimate an answer before you calculate. Something is wrong if your calculated answer is significantly different from your estimate.

Decide When to Use a Calculator

Use a calculator as you need it to find the answer to a question. Do you need to compute, find a graph, solve an equation? There's a good chance the calculator will help. There are many SAT questions when one person will use a calculator and another person will not. There are many other SAT problems when a calculator will not be helpful to anyone. Using a calculator for these problems will just slow you down and cause problems.

Every question on the SAT can be answered without a calculator. You should reach for your calculator only when it will help you.

Calculator Choices

Take the calculator you use in school, or a calculator you are very familiar with. We discuss calculators you might take to the SAT. There are other choices.

TI-84 Plus, TI-83 Plus, or TI-83 Graphing Calculator

The TI-84 Plus is the calculator of choice for most SAT test takers. However, you may buy the TI-83 or TI-83 Plus. You don't need a more advanced calculator, unless that more advanced calculator is your calculator of choice.

Reproduced with permission.

Copyright © Texas Instruments Incorporated.

The TI-83 and the TI-84 have one significant shortcoming. They do not directly display the answer to fraction problems. You have to know how to go through the complicated procedure to see the fraction representation. Direct calculation of fractions is available on the very advanced TI-89 calculator. Direct calculation of fractions is also available on calculators such as the TI-34 Multiview and several other TI scientific calculators.

TI-34 Multiview Scientific Calculator

If you don't want to use a graphing calculator, a scientific calculator such as the TI-34 Multiview will get you through the SAT just fine. The TI-30X IIS or TI-30XA are equally good inexpensive choices.

Reproduced with permission.

Copyright © Texas Instruments Incorporated.

TI-89/TI-Nspire/TI-Nspire CAS

These calculators, which have some more advanced functions, are allowed on the SAT. Be sure you practice with them, since many students find them to be very different from the previous calculators they have used, with many of the menus being different from previous TI calculators. We recommend that you use these only if you're very familiar with them.

Calculators You Can't Use

Here is a list of the types of calculators and devices you can't bring to the SAT.

- You can't bring a computer, even a handheld computer, and you can't bring a calculator with a typewriter keyboard (QWERTY). That means you can't bring the TI-92.
- You can't bring a device that has smartphone like features, such as access to the Internet, wireless, Bluetooth, cellular capability, audio or video record or play, or a camera.
- You can't bring a PDA, a writing pad, or an electronic pen device.
- You can't bring a talking or noisy calculator, one that prints, or one that needs to be plugged in to an electrical outlet.

GEOMETRY FORMULAS AND REFERENCES

Each SAT Mathematics Test shows these geometric formulas and geometry references. You may refer to them when taking the SAT and when working in this book.

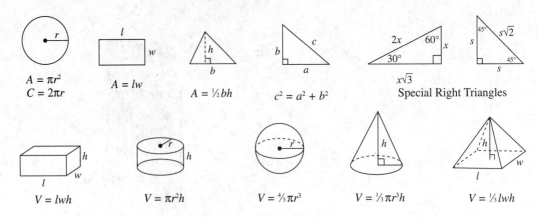

$A = \pi r^2$
$C = 2\pi r$

$A = lw$

$A = \frac{1}{2}bh$

$c^2 = a^2 + b^2$

Special Right Triangles

$V = lwh$

$V = \pi r^2 h$

$V = \frac{4}{3}\pi r^3$

$V = \frac{1}{3}\pi r^3 h$

$V = \frac{1}{3}lwh$

The number of degrees of arc in a circle is 360.

The number of radians of arc in a circle is 2π.

The sum of the measures in degrees of the angles of a triangle is 180.

OK, Let's Do Some Math.

This is the secret. The more math you do, the better you'll get at it. Let's go.

CHAPTER 2

RATIOS, RATES, AND PROPORTIONS

Ratios show the relationship between two numbers. Proportions refer to equal ratios. Rates are ratios that compare two different units, such as miles per hour or beats per minute. Begin with the mathematics review and then complete and correct the practice problems. There are 2 Solved SAT Problems and 21 Practice SAT Questions with answer explanations.

A **ratio** is a comparison of two quantities that can be expressed in three ways.

$$2 \text{ to } 3 \quad \frac{2}{3} \quad 2:3$$

A **proportion** is a statement that two ratios are equal. $\frac{56}{136}$ and $\frac{7}{17}$ form a proportion (are proportional) because $\frac{56}{136} = \frac{\cancel{8} \times 7}{\cancel{8} \times 17} = \frac{7}{17}$. A common technique for checking if two ratios are proportional is to cross multiply. If both cross products are the same, then the two ratios form a proportion, $\frac{56}{136} \diagdown = \diagdown \frac{7}{17}$ $= 136 \times 7 = 952$ $= 56 \times 17 = 952$.

The cross-products are equal so the ratios are equal.

To **solve a proportion,** cross multiply and find the unknown value.

Example 1.

Solve the proportion $\frac{x}{8} = \frac{7}{2}$.

$$\frac{x}{8} = \frac{7}{2} \Rightarrow 2x = 56 \Rightarrow x = 28.$$

Example 2.

Sam's printer prints 4 pages every 15 seconds. How long will it take to print 25 pages?

Write a proportion and solve.

$$\frac{4}{15} = \frac{25}{x} \Rightarrow 4x = 375 \Rightarrow x = 93.75 \text{ sec.}$$

Rate Speed is a common rate that compares distance and time. Use the distance formula.

The distance (D) traveled by an object is equal to the rate (R) at which it is traveling multiplied by the time (T) it has been traveling. $D = R \times T$.

Example 3.

Quinn rode his bike 15 miles at an average rate of 10 miles per hour. For how many minutes did Quinn ride his bike?

Using $D = R \times T$ we get $15 = 10 \times T$. Dividing both sides by 10 gives

$$\frac{15}{10} = T. \quad \Rightarrow \quad T = 1.5 \text{ hours or } 90 \text{ minutes.}$$

Example 4.

Jim is driving home for spring break, a 335-mile trip. He travels an average speed (velocity) of 70 miles per hour for the first 140 miles, and then he travels an average of 65 miles per hour for the last 195 miles. What is Jim's average velocity for the entire trip?

We must first calculate the total time for the trip.

$$D = R \times T \Rightarrow T = \frac{D}{R} \Rightarrow T = \frac{140}{70} + \frac{195}{65} = 2 + 3 = 5 \text{ hours.}$$

Then divide to find the velocity.

$$D = R \times T \Rightarrow R = \frac{D}{T} = \frac{335}{5} = 67 \text{ miles per hour.}$$

Practice Questions

1. A computer store sells 3 computer CDs for $2. How much would 15 computer CDs cost?
2. When Drew goes to the grocery store, for every 2 oranges he buys, he purchases 3 peaches and 1 apple. Drew bought 8 oranges. How many peaches and how many apples did Drew buy?
3. Jason ran for 5 miles. He ran the first 3 miles at an average rate of 8 miles per hour, and then he ran the final 2 miles at an average rate of 6 miles per hour. What is Jason's average speed for the entire 5 miles? (This can be tricky. Refer to the fourth example above for guidance.)

Practice Answers

1. $10.
2. 12 peaches and 4 apples.
3. Approximately 7.2 miles per hour.

SOLVED SAT PROBLEMS

1. For every 50 emails Devon receives, 20 of them are spam. At this rate, if Devon receives 80 emails per day, how many of them will be spam in a given week?

   A. 32
   B. 140
   C. 200
   D. 350

   EXPLANATION: Choice C is correct.

   First, compute the number of spam emails in a day.

   Set up a proportion and simplify if possible:

   $$\frac{20 \text{ spam}}{50 \text{ emails}} = \frac{x \text{ spam}}{80 \text{ emails}} \quad \Rightarrow \quad \frac{2 \text{ spam}}{5 \text{ emails}} = \frac{x \text{ spam}}{80 \text{ emails}}$$

   Cross multiply and solve: $5x = 160 \quad \Rightarrow \quad x = 32$ spam emails

   There are 32 spam emails in a day, so $7(32) = 224$ spam emails in a week.

2. Jordan swam 4 laps. Each lap is 100 meters. Jordan's average speed for the first $\frac{3}{4}$ of his swim was 40 meters per minute, and for the rest of his swim he averaged 35 meters per minute. What is Jordan's average rate for the entire 4 laps, rounded to the nearest whole number?

   EXPLANATION: The correct answer is 39 meters per minute.

   $$D = R \times T \Rightarrow T = \frac{D}{R} \Rightarrow T = \frac{300}{40} + \frac{100}{35} \approx 10.357 \text{ minutes}$$

   Therefore, the 400 meters took approximately 10.357 minutes to complete. Don't round yet.

   Jordan's average rate of speed was

   $$D = R \times T \Rightarrow R = \frac{D}{T} \Rightarrow R = \frac{400}{10.357} \approx 38.621 \text{ meters per minute.}$$

   Round now. That's 39 meters.

RATIOS, RATES, AND PROPORTIONS
PRACTICE SAT QUESTIONS

ANSWER SHEET

Choose the correct answer.
If no choices are given, grid the answers in the section at the bottom of the page.

1. (A)(B)(C)(D)	11. GRID	21. (A)(B)(C)(D)
2. (A)(B)(C)(D)	12. (A)(B)(C)(D)	
3. (A)(B)(C)(D)	13. (A)(B)(C)(D)	
4. GRID	14. (A)(B)(C)(D)	
5. GRID	15. GRID	
6. (A)(B)(C)(D)	16. (A)(B)(C)(D)	
7. (A)(B)(C)(D)	17. (A)(B)(C)(D)	
8. GRID	18. (A)(B)(C)(D)	
9. (A)(B)(C)(D)	19. (A)(B)(C)(D)	
10. (A)(B)(C)(D)	20. (A)(B)(C)(D)	

Use the answer spaces in the grids below if the question requires a grid-in response.

Student-Produced Responses ONLY ANSWERS ENTERED IN THE CIRCLES IN EACH GRID WILL BE SCORED. YOU WILL NOT RECEIVE CREDIT FOR ANYTHING WRITTEN IN THE BOXES ABOVE THE CIRCLES.

4.

5.

8.

11.

15.

PRACTICE SAT QUESTIONS

1. Sound travels approximately 2,200 feet every 2 seconds at sea level. Approximately how many feet will sound travel at sea level in 37 seconds?

   A. 314
   B. 40,700
   C. 81,400
   D. 162,800

2. Asher walks 25 meters in 12.6 seconds. If he walks at the same pace continually, which of these is approximately his distance covered in 2 minutes?

   A. 100 meters
   B. 200 meters
   C. 250 meters
   D. 500 meters

3. It takes 1,000 milliliters of a chemical substance to create the glass for a 700-square-meter window. If a window were 1,400 square meters, about how many windows could 18,000 milliliters of the chemical substance create?

   A. 9 windows
   B. 20 windows
   C. 25 windows
   D. 50 windows

4. If y yards and 55 inches is equal to 595 inches, what is the value of y?

5. A weekly magazine, published 52 weeks a year, sells half-page ads, (2 per page), on one quarter of the 60 pages in each edition. How many half-page ads are published in 12 weeks during June, July, and August?

6. Light travels about 186,000 miles per second. It takes about 4.2 years for light to travel from the star Alpha Centauri to Earth. Which of the following is the best estimate of the distance light will travel during 10% of the time it takes to get from Alpha Centauri to Earth?

   A. 7.12103×10^5 miles
   B. 1.3245120×10^7 miles
   C. 2.46359232×10^8 miles
   D. $2.46359232 \times 10^{12}$ miles

7. A small drone uses half a tank of fuel to fly 30 kilometers. Which of the following is the maximum distance that the drone could fly on $2\frac{1}{2}$ tanks of fuel?

   A. 75 kilometers
   B. 100 kilometers
   C. 125 kilometers
   D. 150 kilometers

8. It takes a long-haul truck driver 3 hours to travel between 2 towns 60 miles apart. What is the best estimate in hours of the time it would take the driver to travel between 2 towns 150 miles apart?

9. The floor of the concession area of the transportation center has 6,700 square feet of glass floor and 1,048 square feet of brick floor. If half the floor of the concession area is occupied by stores and there are 13 stores, what is the average floor area of each store?

   A. 150
   B. 298
   C. 325
   D. 400

10. 1,000 milliliters = 1 liter

    1,000 liters = 1 kiloliter

    The Cryolab stores frozen material in 1-kiloliter containers. The technician adds 400 liters to an empty container. How many milliliters remain empty in the container?

    A. 40
    B. 60
    C. 400,000
    D. 600,000

11. During the Civil War, blockade runners sailed the 28 miles from Fort Fisher to Wilmington, North Carolina, up the Cape Fear River. Today, a modern outboard motor boat can travel that distance in 30 minutes. On average, what would be the speed in miles per hour of an outboard motor boat as it travels this distance?

12. Newspapers are printed on a special kind of paper called newsprint. Newsprint is sold by the ton, and prices go down as the quantity of paper goes up. In one recent year, 20 tons of paper cost $7,600, while 5 tons of paper cost $2,150. Which of the following is closest to the change in cost per ton?

A. $25
B. $50
C. $430
D. $5,450

13.

Number of checkpoints during spaceflight:	13,409
Total number of hours of spaceflight:	1,038 hours
Length of the spacecraft:	327 feet
Number of feet traveled per second in the final portion:	3,800 fps
Total length of time of the final portion:	50 minutes

The research team is reviewing the final portion of a recent spacecraft flight. They have relevant details about the length of the entire flight, the number of electronic checkpoints during the flight, the length of the spacecraft, the number of feet traveled per second in the final portion, and the length of the final portion. About how many feet would the spacecraft travel during the final portion?

A. 988 ft
B. 3,800 ft
C. 190,000 ft
D. 11,400,000 ft

14. A commuter takes 50 minutes to drive a 40-mile trip into the city during the weekday morning rush hour. It takes $\frac{7}{8}$ as long for the commuter to drive home in the afternoon. Driving in on a Saturday takes the commuter $\frac{5}{6}$ times as long as to drive in during the weekday rush. About how much time does he save by traveling in on Saturday than during a weekday?

A. 6 minutes
B. 8 minutes
C. 10 minutes
D. 12 minutes

15. Individual bags of cement at the hardware store weigh 20 pounds and cost $11. The hardware store receives an order for 740 pounds of cement. How many $20 bills is enough to pay for this order?

16. M&M Milk Chocolate Candies are reported to be partitioned as follows: 24% cyan blue, 20% orange, 16% green, 14% bright yellow, 13% red, and 13% brown. If the candies are mixed in a large batch of 1,000,000 that follows the percentages above, how many candies will be orange?

A. 10,000
B. 20,000
C. 100,000
D. 200,000

17. A block of ice is 2 feet by 3 feet by 4 feet. The block melts 0.8 cubic feet each hour. How much ice is left after 3 hours?

A. 0 cubic feet
B. 2.4 cubic feet
C. 21.6 cubic feet
D. 24 cubic feet

18. Polly is making a Halloween costume to dress up as a flower. Her pattern requires 2 feet of fabric for each flower petal. What is the maximum number of petals she can make with 3 yards of fabric? (1 yard = 3 feet.)

A. 2
B. 4
C. 6
D. 8

19. The mason is using flagstones (10 inches on a side) to construct a walk to the front door of a house. She is also using small flat stones (3 inches) on each side. Before construction begins how many of the small square stones laid end to end would match 12 flagstones laid end to end?

A. 10
B. 20
C. 30
D. 40

20. If $\dfrac{3a}{b} = \dfrac{1}{2}$, what is the value of $\dfrac{b}{a}$?

   A. $\dfrac{1}{6}$

   B. $\dfrac{1}{3}$

   C. $\sqrt{6}$

   D. 6

21. At the beginning of registration, 40 freshman students enrolled in Statistics 101, while 50 sophomore students enrolled in the course. If 22 more sophomore students enroll before the end of registration, how many more freshman students must enroll so that $\dfrac{2}{5}$ of the students are freshmen?

   A. 5

   B. 8

   C. 16

   D. 24

▩ EXPLAINED ANSWERS

1. **EXPLANATION: Choice B is correct.**

    In order to solve this equation, you will need to set up a proportion:

    $$\frac{\text{Feet}}{\text{Seconds}} = \frac{2,200}{2} = \frac{x}{37}$$

    Cross multiply: $2x = (37)(2,200)$

    Multiply: $2x = 81,400$

    Divide each side by 2: $\dfrac{2x}{2} = \dfrac{81,400}{2}$

    $x = 40,700$ feet

2. **EXPLANATION: Choice C is correct.**

    Set up a proportion of seconds to meters: $\dfrac{12.6 \text{ seconds}}{25 \text{ meters}} = \dfrac{120 \text{ seconds}}{x \text{ meters}}$, since 120 seconds is 2 minutes.

    Now, cross multiply to solve: $12.6x = 3,000$, and $x = 238$. Standard rounding takes that to 250.

3. **EXPLANATION: Choice A is correct.**

    In order to solve this equation, you will need to set up a proportion to find the amount of ml for each 1,400 sq. m window.

    $$\frac{\text{ml}}{\text{sq. ddm}} = \frac{1,000}{700} = \frac{x}{1,400}$$

    Cross multiply: $700x = (1,000)(1,400)$

    Multiply: $700x = 1,400,000$

    Divide each side by 700: $\dfrac{700x}{700} = \dfrac{1,400,000}{700}$

    Solve the proportion: $x = 2,000$ ml needed for each window

    Then, set up a second proportion to find the number of windows.

    $$\frac{2,000 \text{ ml}}{1 \text{ window}} = \frac{18,000 \text{ ml}}{x \text{ windows}}$$

    Cross multiply: $2,000x = (18,000)(1)$

    Multiply: $2,000x = 18,000$

    Divide each side by 2,000: $\dfrac{2,000x}{2,000} = \dfrac{18,000 \text{ ml}}{2,000 \text{ ml}}$

    Solve the proportion as follows: $x = 9$ windows

4. **EXPLANATION: The correct answer is 15.**

    To solve this problem:

    Write the equation: $y + 55 = 595$

    Subtract: $y = 595 - 55$

    Solve the equation: $y = 540$ inches

    Change the answer to yards: 540 inches $\div 36 = 15$ yards

You can also solve this problem by setting up a ratio: $\dfrac{540 \text{ inches}}{x \text{ yards}} = \dfrac{36 \text{ inches}}{1 \text{ yard}}$

Cross multiply: $\qquad\qquad\qquad 36x = 540$

Divide by 36: $\qquad\qquad\qquad x = 15$ yards

(Note: Most SAT questions will give you a necessary conversion if one is needed, but feet to yards is one they usually expect students to know!)

5. **EXPLANATION: The correct answer is 360.**

Write an equation that translates the problem. In the equation below, 12 is the number of weeks we're talking about, $\dfrac{1}{4}$ is the fraction of the pages devoted to ads, and 60 is the number of pages in each edition. X is the total number of ad pages. $X(\text{ad pages}) = 12\{\frac{1}{4}(60)\}$. Then simplify to get 180. Don't forget, these are half-page ads, though, so each page will have two ads! Multiply by 2 for the final answer, 360 ads.

6. **EXPLANATION: Choice D is correct.**

The proportion needs to compare miles to seconds, so first we have to set up a conversion. Change 4.2 years to seconds. 4.2 years $\times$ 365 days = 1,533 days. 1,533 days $\times$ 24 hours = 36,792 hours. 36,792 hours $\times$ 60 minutes = 2,207,520 minutes; 2,207,520 minutes $\times$ 60 seconds = 132,451,200 seconds. Find 10% of this: 132,451,200 seconds $\times$ 0.10 = 13,245,120 seconds. This value can be used in your proportion.

$$\dfrac{\text{Miles}}{\text{Seconds}} = \dfrac{186{,}000 \text{ miles}}{1 \text{ second}} = \dfrac{x}{13{,}245{,}120 \text{ seconds}}$$

Cross multiply: $\qquad\qquad\qquad X = (186{,}000)(13{,}245{,}120)$

Multiply and solve the proportion: $\qquad X = 2.46359232 \times 10^{12}$ miles

7. **EXPLANATION: Choice D is correct.**

This one is relatively straightforward. All you need to do is compare fuel to kilometers, so write a proportion that does so.

$$\dfrac{\text{Fuel}}{\text{Kilometers}} = \dfrac{0.5}{30} = \dfrac{2.5}{x}$$

Cross multiply: $\qquad\qquad\qquad (0.5)(x) = (2.5)(30)$

Multiply: $\qquad\qquad\qquad\qquad 0.5x = 75$

Divide each side by 0.5: $\qquad\qquad \dfrac{0.5x}{0.5} = \dfrac{75}{0.5}$

Solve the proportion: $\qquad\qquad x = 150$ kilometers

8. **EXPLANATION: The correct answer is 7.5.**

This one is also a straightforward proportion problem, of the type you'd likely see toward the beginning of the math sections.

To solve, write a proportion comparing hours to miles.

$$\dfrac{\text{Hours}}{\text{Miles}} = \dfrac{3}{60} = \dfrac{x}{150}$$

Cross multiply: $\qquad\qquad\qquad (60)(x) = (3)(150)$

Multiply: $\qquad\qquad\qquad\qquad 60x = 450$

Divide each side by 60: $\qquad\qquad \dfrac{60x}{60} = \dfrac{450}{60}$

Solve the proportion: $\qquad\qquad x = 7.5$ hours

9. **EXPLANATION: Choice B is correct.**

   Add to find the total floor area: 6,700 square feet + 1,048 square feet = 7,748 square feet. Divide by 2 to find the floor area of the 13 stores: 7,748 square feet ÷ 2 = 3,874 square feet. Divide this total by 13, the number of stores, to find the average floor area of each store: 3,874 square feet ÷ 13 =298 square feet.

10. **EXPLANATION: Choice D is correct.**

    Change the container amount to liters: 1 kl = 1,000 liters
    Subtract the added 400 liters from the total capacity: 1,000 liters – 400 liters = 600 liters
    Change 600 liters to milliliters: 600 liters × 1,000 = 600,000 ml

11. **EXPLANATION: The correct answer is 56.**

    Write a proportion comparing miles to minutes. Make sure you change the 1 hour to its comparable 60 minutes to fit your proportion.

    $$\frac{\text{Miles}}{\text{Minutes}} = \frac{28}{30} = \frac{x}{60}$$

    Cross multiply: $(30)(x) = (28)(60)$
    Multiply: $30x = 1,680$
    Divide each side by 30: $\dfrac{30x}{30} = \dfrac{1,680}{30}$
    Solve the proportion: $x = 56$ miles per hour

    Or:
    Convert 30 minutes to hours: 30 min = 0.5 hours
    Solve with $D = R \times T$: 28 miles $= R \times 0.5$ hours
    Divide by 0.5 hours: $R = 56$ miles per hour

12. **EXPLANATION: Choice B is correct.**

    Use the ratio cost : tons to discover the unit price for each category. Then find the difference in costs, which will be the increase that the question asks for.

    $$\frac{\text{Cost}}{\text{Tons}} = \frac{2,150}{5} = \frac{x}{1} \text{ or \$430 per ton [cross multiply, so } 5x = 2,150 \text{ and } x = 430]$$

    $$\frac{\text{Cost}}{\text{Tons}} = \frac{7,600}{20} = \frac{y}{1} \text{ or \$380 per ton [cross multiply, so } 20y = 7,600 \text{ and } y = 380]$$

    Subtract $430 – $380 = $50 per ton.

13. **EXPLANATION: Choice D is correct.**

    Use the information from the table to write a proportion comparing feet to seconds. Before you can complete this, you must change the 50 minutes given to a value in seconds—remember to always be sure you're comparing units correctly! In this case, you need to multiply 50 × 60 seconds. This gives us 3,000 as the number of seconds, which is the information needed to complete the proportion.

    $$\frac{\text{Feet}}{\text{Seconds}} = \frac{3,800}{1} = \frac{x}{3,000}$$

    Cross multiply: $(1)(x) = (3,800)(3,000)$
    Multiply and solve the proportion: $x = 11,400,000$ ft

14. **EXPLANATION: Choice B is correct.**

    Eliminate the extra information the problem gives. We only need to compare the travel time on a weekday to the travel time on a Saturday, and the only time we need to think about is driving in, not returning home. We know the weekday travel time is 50 minutes from what the problem gives us. The fraction $\frac{5}{6}$ is important because it compares the Saturday travel time to the weekday time. The travel time on a Saturday is $\frac{5}{6}(50) = 41\frac{2}{3}$. Subtract to find what the problem asks for, which is the difference in the two travel times: $50 - 41\frac{2}{3} = 8\frac{1}{3}$. As often in SAT questions about real-world values, the question asks you to round by using the word "about," so round to 8 minutes.

15. **EXPLANATION: The correct answer is 21.**

    First calculate the total cost of the cement needed.

    $$\frac{\text{Cost}}{\text{Pounds}} = \frac{11}{20} = \frac{x}{740}$$

    Cross multiply: $(20)(x) = (740)(11)$

    Multiply: $20x = 8{,}140$

    Divide each side by 20: $\dfrac{20x}{20} = \dfrac{8{,}140}{20}$

    Solve the proportion: $x = \$407$

    Now divide by 20 to find out how many \$20 bills you will need. $\$407 \div 20 = 20.35$, which means you would need 21 (\$20) bills. You will need to round up to 21 \$20 bills because you cannot have a part of a bill.

16. **EXPLANATION: Choice D is correct.**

    We need only to concentrate on the percentage for the orange candies. Moving the decimal point in the total number to the left one place gives us 10% of the total (100,000). To get the value for 20%, multiply by 2: ($2 \times 100{,}000 = 200{,}000$). You could also change the 20% value to its decimal number (0.2) and multiply by the given total. Multiply $0.2 \times 100{,}000{,}000 = 200{,}000$.

17. **EXPLANATION: Choice C is correct.**

    In order to solve this problem, you will need to use the formula for volume, which can be found on the formulas page at the beginning of each math section. Therefore, you will first calculate the cubic feet of the block of ice by multiplying the 3 dimensions: 2 feet × 3 feet × 4 feet = 24 cubic feet.
        The melting ration is

    $$\frac{\text{Cubic feet}}{\text{Hour}} = \frac{0.8}{1}$$

    Multiply by 3 to get the total amount melting in 3 hours:

    $$3\left(\frac{0.8}{1}\right) = 2.4 \text{ cubic feet}$$

    Subtract to give the amount left: 24.00 cubic feet − 2.40 cubic feet = 21.60 cubic feet.

18. **EXPLANATION: Choice B is correct.**

    For this question, we need to convert yards to feet: 3 yards × 3 feet per yard = 9 feet. Then, set up a proportion of feet to project pieces:

    $$\frac{\text{Feet}}{\text{Project pieces}} = \frac{2}{1}$$

Now set up the proportion: $\dfrac{\text{Feet}}{\text{Project pieces}} = \dfrac{2}{1} = \dfrac{9}{x}$

Cross multiply: $(2)(x) = (9)(1)$

Multiply: $2x = 9$

Divide each side by 2: $\dfrac{2x}{2} = \dfrac{9}{2}$

Solve the proportion: $x = 4.5$ project pieces, but in this case, you have to round down; there's no such thing as part of a piece.

19. **EXPLANATION: Choice D is correct.**

First calculate the length covered by the flagstones. $10 \times 12 = 120$ inches. Now divide this value by the length of each of the small stones. 120 inches $\div 3 = 40$.

20. **EXPLANATION: Choice D is correct.**

In order to solve this proportion, you will need to cross multiply: $(3a)(2) = 1b$. When you multiply the left side of the equation, you'll get: $6a = 1b$. Divide both sides by a, and $6 = \dfrac{b}{a}$.

21. **EXPLANATION: Choice B is correct.**

There were 40 freshmen and 50 sophomores registered at the beginning of registration. Then another 22 sophomore students registered. That is a total of 112 students registered. If x more freshman students must register for $\dfrac{2}{5}$ of the total number of students to be freshmen, the proportion: $\dfrac{\text{Freshman students}}{\text{Total students}} = \dfrac{40 + x}{112 + x} = \dfrac{2}{5}$.

Cross multiply: $5(40 + x) = 2(112 + x)$

Distribute: $200 + 5x = 224 + 2x$

Subtract 200 from both sides of the equation: $5x = 24 + 2x$

Subtract $2x$ from both sides of the equation: $3x = 24$

Solve for x by dividing both sides by 3 to get $x = 8$.

Alternatively, test out the answer choices by adding each choice to 40 and then dividing by the total number of students to try to equal $\dfrac{2}{5}$. When setting up the denominator, remember to add each choice to 112 for the new total.

CHAPTER 3

PERCENTS

A percent shows the ratio of a number to 100. Just finding what percent one number is of another is usually not enough. You will also be asked to find how much a value has increased or decreased, or the new amount of a value after it has increased or decreased. Begin with the mathematics review and then complete and correct the practice problems. There are 2 Solved SAT Problems and 18 Practice SAT Questions with answer explanations.

PERCENT

Percent is a part of a whole represented in hundredths, $\frac{1}{100} = 0.01 = 1\%$. The formula

$$Percent = \frac{Part}{Whole} \Rightarrow Part = Percent \times Whole$$

can be used to solve for the Percent, Part, or Whole.

Example 1.

What percent of 50 is 42?

$$Percent = \frac{Part}{Whole} = \frac{42}{50} = 0.84 = 84\%.$$

Example 2.

30% of 60 is what number?

$$Part = 0.30 \times 60 \Rightarrow Part = 18.$$

Example 3.

15% of what number is 12?

$$12 = 0.15 \times Whole \Rightarrow 80 = Whole.$$

PERCENT INCREASE AND PERCENT DECREASE

$$Percent\ Increase = \frac{Part}{Whole} = \frac{Change\ in\ Amount}{Original\ Amount} = \frac{New\ Amount - Original\ Amount}{Original\ Amount}$$

$$Percent\ Decrease = \frac{Part}{Whole} = \frac{Change\ in\ Amount}{Original\ Amount} = \frac{Original\ Amount - New\ Amount}{Original\ Amount}$$

Example 4.

In a science experiment, 18 ounces of water were placed in a bottle. During the experiment some of the water in the bottle evaporated, leaving 15 ounces of water in the bottle. How much did the water in the bottle change from the original amount?

$$Percent\ Decrease = \frac{18 - 15}{18} = \frac{3}{18} = 0.166\bar{6} = 16.66\bar{6}\% = 16\frac{2}{3}\%$$

Calculator Tip

Here is how to use a graphing calculator to convert $0.166\bar{6}$ to $16\frac{2}{3}\%$.

1. Multiply 0.1666666667 by 100, which equals 16.66666667.
2. Then subtract 16 from 16.66666667, equaling .66666667.
3. Then press the keys that display Ans ▶ Frac on your calculator. The result is $\frac{2}{3}$.
4. Therefore, $0.166\bar{6} = 16.66\bar{6}\% = 16\frac{2}{3}\%$.

   A display of 0.1666666667 means 6 is repeating; 7 is the last digit because the calculator is programmed to round the last digit.

When a value is increasing:

New Amount = (1 + Percent Increase) × Original Amount

When a value is decreasing:

New Amount = (1 − Percent Decrease) × Original Amount

Example 5.

A certain stock started the day at $35 a share. By the end of the day the stock had increased by 20%. What is the new price of the stock?

New Amount = (1 + 0.2) × 35 = 1.2 × 35 = $42.

Practice Questions

1. Jack is 30 years old and Alice is 35 years old. What percent of Alice's age is Jack, rounded to the nearest tenth of a percent?
2. 90% of the students in a high school have a cell phone. If there are 750 students in the school, how many of the students have cell phones?
3. Richard's car costs 70% of Brian's car. If Richard's car cost $17,500, what is the price of Brian's car?
4. A jacket went on sale. The original price was $24, and the sale price was $18. At what percent was the jacket discounted?
5. Over the summer Jason grew from 70 inches to 73.5 inches. What percent has Jason's height increased?
6. Last season a certain basketball player averaged 25 points a game. This season the basketball player increased her average by 12%. What is the basketball player's new scoring average?

Practice Answers

1. 85.7%.
2. 675 students.
3. $25,000.
4. 25%.
5. 5%.
6. 28 points per game.

SOLVED SAT PROBLEMS

1. In a 20-game season a certain basketball player scores at least 15 points in 60% of the first 10 games, 40% of the next 5 games, and 20% in the last 5 games. In what percent of all 20 games did this player score less than 15 points?

   A. 40%
   B. 45%
   C. 50%
   D. 55%

   EXPLANATION: Choice D is correct.

   Multiply and add to find the number of games in which the player scored at least 15 points.

   $0.6 \cdot 10 + 0.4 \cdot 5 + 0.2 \cdot 5 = 9$

   The player scored at least 15 points in 9 games, which means that the player scored less than 15 points in $20 - 9 = 11$ games.

   $\frac{11}{20} = 0.55 = 55\%$ of the games the player scored less than 15 points.

2. The cost of grapes at a grocery store is normally $2.99 a pound. There was a sale on grapes for 65% off the normal price. What is the sale price of the grapes to the nearest cent?

   A. $1.05
   B. $2.05
   C. $2.99
   D. $3.05

   EXPLANATION: Choice A is correct.

   Since the grapes are on sale for 65% off the original price, the new price will be $100\% - 65\% = 35\%$ off the original price.

   Therefore, the sale price will be $2.99 (0.35) = $1.0465. Rounded to the nearest cent is $1.05 per pound.

PERCENTS
PRACTICE SAT QUESTIONS

ANSWER SHEET

Choose the correct answer.
If no choices are given, grid the answers in the section at the bottom of the page.

1. Ⓐ Ⓑ Ⓒ Ⓓ
2. Ⓐ Ⓑ Ⓒ Ⓓ
3. Ⓐ Ⓑ Ⓒ Ⓓ
4. Ⓐ Ⓑ Ⓒ Ⓓ
5. Ⓐ Ⓑ Ⓒ Ⓓ
6. Ⓐ Ⓑ Ⓒ Ⓓ
7. Ⓐ Ⓑ Ⓒ Ⓓ
8. Ⓐ Ⓑ Ⓒ Ⓓ
9. Ⓐ Ⓑ Ⓒ Ⓓ
10. Ⓐ Ⓑ Ⓒ Ⓓ

11. Ⓐ Ⓑ Ⓒ Ⓓ
12. Ⓐ Ⓑ Ⓒ Ⓓ
13. **GRID**
14. **GRID**
15. Ⓐ Ⓑ Ⓒ Ⓓ
16. Ⓐ Ⓑ Ⓒ Ⓓ
17. Ⓐ Ⓑ Ⓒ Ⓓ
18. Ⓐ Ⓑ Ⓒ Ⓓ

Use the answer spaces in the grids below if the question requires a grid-in response.

| Student-Produced Responses | ONLY ANSWERS ENTERED IN THE CIRCLES IN EACH GRID WILL BE SCORED. YOU WILL NOT RECEIVE CREDIT FOR ANYTHING WRITTEN IN THE BOXES ABOVE THE CIRCLES. |

PRACTICE SAT QUESTIONS

1. A steel company uses metric tons and short tons to weigh its products. A short ton weighs approximately 9% less than a metric ton. If one steel shipment weighs 43 short tons, what is the weight of that shipment in metric tons?

   A. 3.87 metric tons
   B. 34 metric tons
   C. 39.13 metric tons
   D. 40.23 metric tons

2. At the Pismey glass factory, there are two production lines for lightbulbs. One production line creates 50-watt bulbs, while the other production line creates 100-watt bulbs. On average, inspectors on the 100-watt line find that 5% of the bulbs are flawed, while inspectors on the 50-watt line find 8% flawed bulbs. In one hour the 100-watt line produces 840 bulbs, while the 50-watt line produces 1,050 bulbs. On average, what would be the total number of flawed bulbs per hour?

   A. 42
   B. 84
   C. 126
   D. 130

3. Common Ways of Commuting to Work

Car, truck, van	84%
Public transportation	6%
Bike	1%
Walk	3%
Motorcycle	2%
Home office	4%

   The table above shows the results of responses to a survey about the most popular ways for people to commute to work or if the person has a home office. If the survey was based on 2,000 people, how many people commuted to work by public transportation or walking?

   A. 40
   B. 60
   C. 120
   D. 180

4. The manager of a grocery store assigns all part-time hours to various store functions. In one month, 45% of part-time hours were allocated to restocking, 25% of part-time hours were allocated to maintenance, and 20% of part-time hours were assigned to receiving. The manager assigned the remaining 20 hours for general store work. In that month, how many more hours were assigned to restocking than to receiving?

   A. 20
   B. 50
   C. 75
   D. 80

5. A clothing store adds 10% to the price of its most popular items and charges 7% sales tax. What is the final amount charged by the store if the original price of the item was P?

   A. $1.07P$
   B. $1.177P$
   C. $2.07P$
   D. $(.07)(1.1)P$

6. An engineer tracks electricity prices (in dollars) and notes the pattern shown below.

JUNE	JULY	AUGUST
628	827	535

   Which of the following is closest to the percent change between JUNE and AUGUST?

   A. 3% decrease
   B. 15% decrease
   C. 17% decrease
   D. 85% decrease

7. A landscaper adds a 9% fuel surcharge to a customer's bill. If the normal fuel charge was $26, what was the cost of fuel after the surcharge was added?

   A. $23.66
   B. $28.34
   C. $28.60
   D. $35.00

8. An arctic weather observer measured snow depths at different places around the isolated base. She noticed that the snow depth in the valley was always 18% more than the snow depth on the mountainside. In the most recent observation, there were 19 feet of snow on the mountainside. What is the likely depth of the snow in the valley?

   A. 15.58 ft
   B. 19.18 ft
   C. 22.42 ft
   D. 34.2 ft

9. The product of two numbers is 1,452. One of the numbers is 200% larger than the other number. What is the sum of the numbers?

   A. 22
   B. 66
   C. 88
   D. 363

10. The holder of a cash-back credit card gets "5% cash back" for each purchase made with the card. During one month, the cash-back amount is increased by an additional 20% back for supermarket purchases. In that month, what is the least amount, in whole dollars, of supermarket purchases that would yield a cash-back amount over $25?

    A. $100
    B. $101
    C. $125
    D. $200

11. During a month of four 40-hour workweeks, the computer programmer was not working on his computer 30% of the time. How many minutes during the four 40-hour workweeks was the programmer not working on a computer?

    A. 48
    B. 480
    C. 2,880
    D. 2,900

12. The mall manager noted a yearly decrease in hourly mall traffic. The manager commissioned a study that resulted in the table shown below. If the percent of decrease from 2020 to 2021 increased by 10 percentage points in the following year, what would be the hourly mall traffic number for 2022? (Give your answer rounded to the nearest whole number.)

    Hourly Mall Traffic

2020	18,600
2021	17,298
2022	

    A. 14,357
    B. 15,568
    C. 15,996
    D. 16,087

13. A cement plant has 20 tons of sand gravel mix that is 15% gravel and 85% sand. How many tons of gravel would have to be added to the sand gravel mix to create a mix of 20% gravel?

14. Jeff bought a collectible video game that cost $300. Each year, the value of the game, according to the leading collectors in the video game industry, will increase 10% until at 10 years after release, it reaches its maximum value. What number, rounded to the nearest hundredths, would you multiply 300 by to determine the value of the game after four years?

15. After the addition of 10% sales tax, Ada paid $44.00 for her new jacket. What was the price of the jacket before tax?

    A. $34.00
    B. $40.00
    C. $48.40
    D. $48.80

16. Diane buys two brands of snack mix to serve to her son's basketball team at the end-of-season banquet and will combine them into one bowl for serving. Brand A contains 65% peanuts by weight. Brand B contains 35% peanuts by weight. In all, they contain 25 pounds of peanuts as part of 45 pounds of total snack mix. Which equation best represents this scenario? In the choices below, g is the number of pounds of Brand A and h is the number of pounds of Brand B.

    A. $0.65g + 0.35h = 25$
    B. $0.35g + 0.65h = 25$
    C. $0.35g + 0.65h = 45$
    D. $0.65g + 0.35h = 45$

17. The weight of an object on Asteroid Alpha is half of its weight on Earth, while its weight on Asteroid Beta is 70% of its weight on Earth. Astronaut Bob's space helmet weighs 50 pounds on Earth. Approximately how many more pounds does it weigh on Asteroid Beta than on Asteroid Alpha?

    A. 10 pounds
    B. 25 pounds
    C. 35 pounds
    D. 100 pounds

18. How many liters of a punch that is 10% pineapple juice must be added to 6 liters of a punch that is 20% orange juice to obtain a punch that is 15% juice? Assume that no other types of juice are part of the punch.

    A. 2
    B. 4
    C. 6
    D. 12

![EXPLAINED ANSWERS]

EXPLAINED ANSWERS

1. **EXPLANATION: Choice C is correct.**

   Change 43 short tons to its metric ton value. A metric ton weighs more than a short ton so there will be fewer than 43 metric tons. First subtract 9% from 100% (100% − 9% = 91%). Then multiply 43 short tons by 0,91 = 39.13 metric tons.

2. **EXPLANATION: Choice C is correct.**

   Organize your data and then multiply the decimal equivalents for flawed bulbs times total bulbs in each group. $(0.08 \times 1{,}050 = 84)$; $(0.05 \times 840 = 42)$; add the group totals $(84 + 42 = 126)$.

3. **EXPLANATION: Choice D is correct.**

   Add the percentages for both of the stated categories (6% + 3% = 9%). Multiply the decimal equivalent for 9% times the total number of people $(0.09 \times 2{,}000 = 180$ people$)$.

4. **EXPLANATION: Choice B is correct.**

   First we need to pay attention that the facts are given in two distinct forms (percents and numeric values). We can find the total number by concentrating on the 20 hours assigned to general store work and the remaining facts given as percentages. Adding all the given categories that are given as percentages, we get 45% + 25% + 20% = 90%. We then know that the 20 hours assigned to general store work can only be equivalent to 10% of the total hours assigned. Write an equation that states this relationship: 10% of X = 20, or $0.1x = 20$. Divide 20 by 0.1 to get 200, which gives the total numeric hours. Now we can use this to calculate the restocking number $(0.45 \times 200 = 90)$ and the receiving number $(0.2 \times 200 = 40)$. Subtract these values, $90 − 40 = 50$, to find the answer to this problem: 50 more hours were assigned to restocking than to receiving.

5. **EXPLANATION: Choice B is correct.**

   If P is the original amount of the item, then we must increase this first by 0.1, which is equivalent to 10% $(P + 0.1P = 1.1P)$. This amount $(1.1P)$ then must be increased by the 7% tax $(0.07)(1.1P) = 0.077P$. This tax must then be added to the 10% increase price $(0.077P + 1.1P = 1.177P)$.

   Alternatively, we can pick a value for P to avoid working with the variable. If the original price, P, is \$100, then adding 10% to the price is $0.1 \times 100 = \$10$ additional. The new price is $\$100 + \$10 = \$110$. Calculate 7% sales tax and add to the price: $0.07 \times \$110 = \$7.7$; $\$7.7 + \$110 = \$117.70$. Now, plug in $P = 100$ into all answer choices to see which one gives \$117.70. B is the only answer choice to do so.

6. **EXPLANATION: Choice B is correct.**

   The problem states only the comparison between June and August. This is a decrease amount. To find the percentage of decrease, build a ratio of change $(628 − 535 = 93)$: original (628). Divide and change to a % $(93 \div 628 = 14.8\%)$. 15% decrease is the best answer.

7. **EXPLANATION: Choice B is correct.**

   Calculate the surcharge $(0.09 \times \$26.00 = \$2.34)$. Add it to the original cost $(\$2.34 + \$26.00 = \$28.34)$.

8. **EXPLANATION: Choice C is correct.**

   Find 18% of 19 by multiplying by 0.18 $(0.18 \times 19 = 3.42)$. Then, add that number to the original 19 feet to get 22.42 feet.

9. **EXPLANATION: Choice C is correct.**

   The problem states that there are two distinct numbers (x) and (y). It also states that the larger number $(y) = x + 200\%x$. This means $y = x + 2x$, or $3x$. Use these values for the given product. $(x)(3x) = 1{,}452$. Solve for $x = (3x^2) = 1{,}452$; $3x^2 = 1{,}452$; $x^2 = 484$; $x = 22$, thus $y = 3x$ or 66. This makes the sum $22 + 66 = 88$.

10. **EXPLANATION: Choice B is correct.**

Find the percentage for supermarket purchases that month (20 + 5 = 25). "Over" means "greater than." Write an inequality using the given facts (25% × x > 25). Solve for x (0.25x > 25; x > 25 ÷ 0.25; x > 100). Answer B is the least amount in whole dollars that yields a cash-back amount over $25.

11. **EXPLANATION: Choice C is correct.**

Find the total number of hours worked (4 × 40 = 160 hours). Next calculate 30% of this total (0.3 × 160 = 48 hours). Now change this value to minutes (48 hours × 60 minutes = 2,880 minutes).

12. **EXPLANATION: Choice A is correct.**

Find the percentage of decrease from 2020 to 2021: change (18,600 − 17,298 = 1,302): original (18,600). Divide 1,302 ÷ 18,600 = 0.07 = 7%. Increase this number by 10% (7% + 10% = 17%). Solve for the unknown: change (x): original (17,298 = 17%); 17,298 × 0.17 = 2,940.66; round to the nearest whole number (2,941). Subtract from the 2021 original amount (17,298 − 2,941 = 14,357).

13. **EXPLANATION: The correct answer is 1.25.**

Solve this by setting up an equation to model the situation. You can ignore the percentage of sand: what we care about is the percentage of gravel. The equation is 0.15(20) to represent the amount of gravel that currently exists in the mix, and you're adding something that is 100% gravel in an unknown amount (1x), to equal a mixture that will be 0.2(20 + x). The expression in the parentheses represents the new total amount of the mixture (the original 20 tons plus however much gravel you are adding). You can then set up the equation and solve for x as follows:

$0.15(20) + 1x = 0.2(20 + x)$

$3 + x = 4 + 0.2x$

$x = 1 + 0.2x$

$0.8x = 1$

$x = 1.25$, so the answer is 1.25 tons of gravel.

14. **EXPLANATION: The correct answer is 1.46.**

The increase of 10% each year is the equivalent of multiplying the original price of the game by 1.1 (it maintains its value and grows 10%, which is the 0.1). So, after 1 year, the value of the game is $300(1.1), and after two years, it would be $300(1.1)(1.1). After four years, the value should be $300(1.1)(1.1)(1.1)(1.1) = 300 × 1.4641, so the answer is 1.4641, which rounds to 1.46.

15. **EXPLANATION: Choice B is correct.**

Work backwards from the answer choices to solve this question! Start with the middle answer choice for the most efficiency. If the answer is $48.40, then the price of the dress with the tax would have to be even higher; therefore, you can eliminate both C and D. Now, try either A or B. For answer choice B, $40.00 × 0.1 = $4.00, and $40.00 + $4.00 = $44.00, which is what the problem says we're searching for. That's therefore the correct response.

Alternatively (or if this question is presented as a grid-in), you can set up an equation. The addition of a 10% sales tax is like multiplying the original price by 1.1, so 1.1x = $44.00. Divide both sides by 1.1, and x = $40.00.

16. **EXPLANATION: Choice A is correct.**

Percents can be represented by decimals or fractions—in this case, the answers to the problem are in decimals. Brand A has 65% peanuts: represent that as 0.65g, and Brand B is 35% peanuts, or 0.35h. All the peanuts weigh 25 pounds, on the right side of the equation, so A is the answer choice that fulfills these conditions. Choices C and D are incorrect because the equations show the total weight of all the nuts on the right side, which is in there as extra information or a distraction—not something you need for the problem.

17. **EXPLANATION: Choice A is correct.**

You can solve this in several steps: 0.5 (50%) of x (the helmet's weight on Earth) will be its weight on Asteroid Alpha, and $0.7x$ will therefore be the weight on Asteroid Beta. Subtract to solve the problem. Set the helmet's weight equal to 50 pounds:

0.5(50) = 25 (weight on Asteroid Alpha).

0.7(50) = 35 (weight on Asteroid Beta).

35 − 25 = 10 pounds more on Asteroid Beta than on Asteroid Alpha.

18. **EXPLANATION: Choice C is correct.**

In order to solve this question, you will first need to write the following equation: 6(20%) + x(10%) = $(x + 6)$(15%). Then change the percents to decimals: 6(0.20) + x(0.10) = $(x + 6)$(0.15). Multiplying 6 by 0.20 on the left side of the equation and 6 by 0.15 on the right side of the equation will yield: $1.2 + 0.1x = 0.15x + 0.9$. Then subtract 0.9 from each side: $0.3 + 0.1x = 0.15x$. Then subtract $0.1x$ from each side: $0.3 = 0.05x$. Finally, solve for x: $\dfrac{0.3}{.05} = x$. Write the solution: $x = 6$ liters. Therefore, you will need to add 6 liters of 10% pineapple juice punch.

CHAPTER 4

SOLVING EQUATIONS

This section shows you how to evaluate expressions and how to solve equations with a single variable. An **expression** is a variable, a number, or combination of variables, numbers, and operation symbols. Here are some examples of expressions: 4, y, $x + 3$, $x + y$, xy. To evaluate an expression, to substitute values for variables and complete the operations. An **equation** uses an equal sign to show that two expressions have the same value. Begin with the mathematics review and then complete and correct the practice problems. There are 2 Solved SAT Problems and 21 Practice SAT Questions with answer explanations.

SOLVING EQUATIONS

Solve the following equations. Isolate the variable and solve for x.

Example 1.

Solve for x in the equation $45 = 13 + 4x$

Subtract 13 from both sides of equation: $32 = 4x$

Divide both sides of equation by 4: $8 = x$

Example 2.

Solve for x in the equation $5 + \frac{1}{3}x = 9$.

Subtract 5 from both sides of equation: $\frac{1}{3}x = 4$

Multiply both sides of equation by 3: $x = 12$

Example 3.

Solve for x in the equation $\frac{3}{5}x = \frac{5}{4}$.

Multiply both sides of equation by $\frac{5}{3}$: $x = \frac{5}{4} \times \frac{5}{3}$

Simplify solution: $x = \frac{25}{12}$ or $2\frac{1}{12}$

Example 4.

Solve for x in the equation $15x - 9 = 8x + 12$.

Add 9 to both sides of equation: $15x = 8x + 21$

Subtract $8x$ from both sides of equation: $7x = 21$

Divide both sides of equation by 7: $x = 3$

Example 5.

Solve for x in the equation $3(4x - 7) - 2(5x - 9) = 11$.

Distribute on the left side of the equation: $12x - 21 - 10x + 18 = 11$
Combine like terms on the left side of the equation: $2x - 3 = 11$
Add 3 to both sides of the equation: $2x = 14$
Divide both sides of the equation by 2: $x = 7$

Example 6.

Solve for x in the equation $\dfrac{x - 3}{x + 7} = 6$.

Multiply both sides of the equation by $x + 7$: $x - 3 = 6(x + 7)$
Distribute the right side of the equation: $x - 3 = 6x + 42$
Add 3 to both sides of the equation: $x = 6x + 45$
Subtract $6x$ from both sides of the equation: $-5x = 45$
Divide both sides of the equation by -5: $x = -9$

Example 7.

Solve for x in the equation $\dfrac{x + 2}{x - 4} = \dfrac{7}{9}$.

Cross multiply: $9(x + 2) = 7(x - 4)$
Distribute both sides of the equation: $9x + 18 = 7x - 28$
Subtract 18 from both sides of the equation: $9x = 7x - 46$
Subtract $7x$ from both sides of the equation: $2x = -46$
Divide both sides of the equation by 2: $x = -23$

Example 8.

Solve for $x + 7$ in the equation $5(x + 7) = 30$.

The question is asking for $x + 7$, so isolate $x + 7$, not just x
Divide both sides of the equation by 5: $x + 7 = 6$

Example 9.

Solve for $x - 3$ in the equation $4x - 12 = 12$.

The question is asking for $x - 3$, so isolate $x - 3$
Factor out a 4 on the left side of equation: $4(x - 3) = 12$
Divide both sides of equation by 4: $x - 3 = 3$

Example 10

Solve for $10x + 2$ in the equation $5x - 9 = 11$.

The question is asking to solve for $10x + 2$. Unlike the previous two problems, it is not simple to isolate $10x + 2$, but it still can be done.

Add 9 to both sides of the equation: $5x = 20$

Multiply both sides of the equation by 2: $10x = 40$

Add 2 to both sides of the equation: $10x + 2 = 42$

As an alternate method that may often be easier, you could always solve for x first and then find $10x + 2$.

Add 9 to both sides of the equation: $5x = 20$

Divide both sides of the equation by 5: $x = 4$

Substitute $x = 4$ into $10x + 2$: $10x + 2 = 10(4) + 2 = 40 + 2 = 42$.

Example 11.

11 times the number x added to 3 equals 6 times the number x subtracted by 12. Find x.

Write an equation from the given information: $11x + 3 = 6x - 12$

Subtract $6x$ from both sides of the equation: $5x + 3 = -12$

Subtract 3 from both sides of the equation: $5x = -15$

Divide both sides of the equation by 5: $x = -3$.

Practice Questions

1. Solve for x in the equation $7 + \dfrac{2}{5}x = 21$.

2. Solve for x in the equation $5(2x + 3) - 4(6x + 3) = 4$.

3. Solve for x in the equation $\dfrac{x + 5}{x - 3} = 5$.

4. Solve for $x + 2$ in the equation $3x + 6 = 18$.

5. 8 times the number x subtracted by 6 equals 3 times the number x added to 4. Find x.

Practice Answers

1. $x = 35$.

2. $x = -\dfrac{1}{14}$.

3. $x = 5$.

4. $x + 2 = 6$.

5. $x = 2$.

SOLVED SAT PROBLEMS

1. $4(5x - 3) - 3(2x + 1) = 6(x - 2)$

   What value of x is the solution of the equation above?

   EXPLANATION: The correct answer is $\dfrac{3}{8}$ or 0.375.

   Distribute both sides of equation: $20x - 12 - 6x - 3 = 6x - 12$

   Combine like terms on left side of equation: $14x - 15 = 6x - 12$

   Add 15 to both sides of equation: $14x = 6x + 3$

   Subtract $6x$ from both sides of equation: $8x = 3$

   Divide both sides of equation by 8: $x = \dfrac{3}{8} = .375$

2. $\dfrac{2}{3}x - \dfrac{3}{4} = \dfrac{5}{12}x - \dfrac{1}{8}$

   What value of x is the solution of the equation above?

   EXPLANATION: The correct answer is $\dfrac{1}{2}$ or 0.5.

   Find the least common denominator (LCD): 24

   Multiply both sides of equation by LCD = 24:

   $$24 \times \left(\dfrac{2}{3}x - \dfrac{3}{4} \right) = 24 \times \left(\dfrac{5}{12}x - \dfrac{1}{8} \right)$$

   Distribute the 24 on both sides of equation:

   $$24 \times \left(\dfrac{2}{3}x \right) - 24 \times \left(\dfrac{3}{4} \right) = 24 \times \left(\dfrac{5}{12}x \right) - 24 \times \left(\dfrac{1}{8} \right)$$

   Simplify: $8 \times \left(\dfrac{2}{1}x \right) - 6 \times \left(\dfrac{3}{1} \right) = 2 \times \left(\dfrac{5}{1}x \right) - 3 \times \left(\dfrac{1}{1} \right) = 16x - 18 = 10x - 3$

   Add 18 to both sides of equation: $16x = 10x + 15$

   Subtract $10x$ from both sides of equation: $6x = 3$

   Divide both sides of equation by 6: $x = \dfrac{3}{6} \quad \rightarrow \quad x = \dfrac{1}{2} = 0.5$

SOLVING EQUATIONS
PRACTICE SAT QUESTIONS

ANSWER SHEET

Choose the correct answer.
If no choices are given, grid the answers in the section at the bottom of the page.

1. (A) (B) (C) (D)   11. (A) (B) (C) (D)   21. (A) (B) (C) (D)
2. (A) (B) (C) (D)   12. (A) (B) (C) (D)
3. (A) (B) (C) (D)   13. GRID
4. GRID   14. GRID
5. (A) (B) (C) (D)   15. (A) (B) (C) (D)
6. (A) (B) (C) (D)   16. (A) (B) (C) (D)
7. (A) (B) (C) (D)   17. (A) (B) (C) (D)
8. (A) (B) (C) (D)   18. (A) (B) (C) (D)
9. GRID   19. (A) (B) (C) (D)
10. (A) (B) (C) (D)   20. (A) (B) (C) (D)

Use the answer spaces in the grids below if the question requires a grid-in response.

Student-Produced Responses ONLY ANSWERS ENTERED IN THE CIRCLES IN EACH GRID WILL BE SCORED. YOU WILL NOT RECEIVE CREDIT FOR ANYTHING WRITTEN IN THE BOXES ABOVE THE CIRCLES.

4.

9.

13.

14.

PRACTICE SAT QUESTIONS

1. $x - 2x + 3 + x - 4 = x + 2x - 6$.

    Given that the above equation is true, what is the value of x?

    A. $\dfrac{-3}{5}$

    B. $\dfrac{3}{5}$

    C. $\dfrac{-5}{3}$

    D. $\dfrac{5}{3}$

2. On Tuesday, Kevin live-tweeted a TV show for his followers. He tweeted u times per hour for 2 hours. His friend David made comments on his tweets and replied w times per hour for 4 hours. Which of the following represents the total number of tweets and replies sent by Kevin and David on Tuesday?

    A. $6uw$
    B. $8uw$
    C. $2u + 4w$
    D. $4u + 2w$

3. Compound interest can be calculated using the formula:

    $A = P\left(1 + \dfrac{r}{n}\right)^{(nt)}$, where the principal amount is

    (P), the annual interest rate is (r) as a decimal, the time factor is (t), and the number of compound periods is (n). Which of the following gives P in terms of A, r, n, and t?

    A. $P = A - \left(1 + \dfrac{r}{n}\right)$

    B. $P = A - \left(1 + \dfrac{r}{n}\right)^{(nt)}$

    C. $P = A / \left(1 + \dfrac{r}{n}\right)^{(nt)}$

    D. $P = \left(1 + \dfrac{r}{n}\right)^{(nt)} / A$

4. $\dfrac{2}{3}x - 0.5B = 0$

    If $x = 3$ in the equation above, what is the value of B?

5. If $\dfrac{4}{y + 3} = h$ and $h = 2$, what is the value of y?

    A. -1
    B. 0
    C. 1
    D. 2

6. $\dfrac{1}{5}m = \dfrac{6}{3}$

    Solve the equation for m.

    A. 10
    B. 20
    C. 30
    D. 40

7. $\dfrac{4}{5} = \dfrac{2}{5}q$

    Find the value of q.

    A. 2
    B. 4
    C. 6
    D. 8

8. What is $s - 9$ if $4s - 6 = 34$?

    A. -1
    B. 1
    C. 7
    D. 10

9. $\dfrac{d}{4k} = \dfrac{2}{3}$. What is the value of $\dfrac{k}{d}$?

10. If $8(x + y) = 64$, what is the value of $x + y$?

    A. 8
    B. 10
    C. 12
    D. 14

11. If $\dfrac{3}{x} = \dfrac{9}{x + 10}$, what is the value of $\dfrac{x}{5}$?

    A. 1
    B. 2
    C. 5
    D. 6

12. If $5x$ subtracted from 12 is 6 less than 13, what is the value of x?

    A. -1
    B. 0
    C. 1
    D. 2

13.

    A   B   C   D

In the line above, $AB = BC$. If the length of AC is $2x + 1$, the length of CD is $3x - 1$, and $AD = 75$, what is the length of BD? Round your answer to the nearest tenth.

14. Find the value of x.

$$\frac{5}{6}x - \frac{1}{3}x = \frac{3}{4} - \frac{1}{6}$$

15. Which of the following is equivalent to $8(x + 2) - 5$?

    A. $8x + 11$
    B. $8x - 1$
    C. $8x - 9$
    D. $8x - 5$

16. If $a - b = 4$ and $\frac{b}{2} = 12$, how much is $a + b$?

    A. 24
    B. 28
    C. 48
    D. 52

17. A mathematician is working on a problem with two functions, $f(x)$ and $g(x)$. In the problem $f(x) - g(x) = 4$. Which result below reflects that relationship?

    A. $f(x) = 4 - g(x)$
    B. $f(x) = 4 + g(x)$
    C. $f(x) = 4g(x)$
    D. $f(x) = \frac{1}{4}g(x)$

18. $f(x) = 20 + 2.5y$

Advertisements are placed in a newspaper according to the equation seen above. When an ad with length of y words appears in the paper, the price of the ad is $f(x)$. What is the number of words in an ad that costs \$70?

    A. 20
    B. 50
    C. 70
    D. 125

19. A computer programmer uses the formula $p = 9wh$ to determine the number of pixels, p, in a picture that is w inches wide and h inches high, at his preferred graphic resolution. Which of the following is a valid way to express h?

    A. $h = 9pw$
    B. $h = \dfrac{9}{pw}$
    C. $h = \dfrac{p}{9w}$
    D. $h = \dfrac{w}{9p}$

20. When 15 is added to 3 times an integer value h, 6 is the answer. What integer value do you get if you add 10 to 2 times h?

    A. -10
    B. -4
    C. -1
    D. 4

21. $\dfrac{x - 2}{4} = n.$ $n = 5.$ What is $x + 5$?

    A. $\dfrac{3}{4}$
    B. 7
    C. 22
    D. 27

EXPLAINED ANSWERS

1. **EXPLANATION: Choice D is correct.**

   Combine like terms on each side of the equation: $-1 = 3x - 6$. Then add 6 to both sides to get: $5 = 3x$. Finally, divide both sides by 3: $\frac{5}{3} = \frac{3x}{3}$. Therefore, $\frac{5}{3} = x$.

2. **EXPLANATION: Choice C is correct.**

   Although you can plug in numbers to solve this problem, as with all problems with variables in the questions and answer choices, this one is fairly easy to solve with logic: Kevin tweets u times per hour for 2 hours, so $2u$. Similarly, David must be represented by $4w$.

3. **EXPLANATION: Choice C is correct.**

   The objective here is to isolate P and give its value equal to all the other variables. Because P is being multiplied by the other variables, to isolate it you must divide both sides of the equation by the other factor $\left(1 + \frac{r}{n}\right)^{(nt)}$. This produces $A / \left(1 + \frac{r}{n}\right)^{(nt)} = P$. This is the same as answer C.

4. **EXPLANATION: The correct answer is 4.**

   Substitute 3 for x in the equation: $\left[\frac{2}{3}\left(\frac{3}{1}\right)\right] - 0.5B = 0$. Simplify the left side: $2 - 0.5B = 0$. Subtract 2: $-0.5B = -2$. Divide both sides by -0.5: $\frac{-0.5b}{-0.5} = \frac{-2}{-0.5}$. Solve: $B = 4$.

5. **EXPLANATION: Choice A is correct.**

   To solve this problem, you will need to set up the equation by replacing the h with the given number, 2: $\frac{4}{y + 3} = 2$. Then, cross multiply to get: $2(y + 3) = 4$. Distribute the coefficient to get: $2y + 6 = 4$. Subtract 6 to each side to get: $2y = -2$. Divide 2 on each side; therefore: $Y = -1$.

6. **EXPLANATION: Choice A is correct.**

   In order to solve this equation, you will first need to multiply both sides by 5 to get rid of the fraction, which gives: $m = \frac{30}{3}$. Then divide by 3. Therefore, $m = 10$.

7. **EXPLANATION: Choice A is correct.**

   Start by multiplying both sides by $\frac{5}{2}$ to get rid of the fraction, which gives you: $q = \frac{5}{2}\left(\frac{4}{5}\right)$. Then multiply to get: $q = \frac{20}{10}$. Finally, simplify by 10; therefore, $q = 2$.

8. **EXPLANATION: Choice B is correct.**

   Solve the equation given. Add 6 to each side: $4s = 40$. Divide by 4: $s = 10$. Now subtract 9 to find the answer to the problem.

9. **EXPLANATION: The correct answer is $\frac{3}{8}$, or 0.375.**

Using the first fraction, $d = 2$ and $4k = 3$. Solve for k to get $k = \frac{3}{4}$. The answer is thus $\frac{\frac{3}{4}}{2}$, which is $\frac{3}{4} \times \frac{1}{2} = \frac{3}{8}$, or 0.375.

Alternatively, cross multiply: $3d = (4k)(2)$. Simplify: $3d = 8k$. Now rearrange for what the question asks for $\left(\frac{k}{d}\right)$ by dividing both sides by $8d$: $\frac{(3d)}{(8d)} = \frac{(8k)}{(8d)}$, or $\frac{3}{8} = \frac{k}{d}$.

10. **EXPLANATION: Choice A is correct.**

Isolate the $(x + y)$ term by dividing both sides by 8 to find the answer: 8.

11. **EXPLANATION: Choice A is correct.**

Start by cross multiplying: $9x = 3(x + 10)$. Distribute the coefficient: $9x = 3x + 30$. Then subtract $3x$ from both sides, so you get: $6x = 30$. Simplify by dividing 6 from each side and solve: $\frac{6x}{6} = \frac{30}{6}$. Solve: $x = 5$. Finally, plug 5 back into the question to get: $\frac{x}{5} = \frac{5}{5} = 1$.

12. **EXPLANATION: Choice C is correct.**

Start by writing the equation: $13 - 6 = 12 - 5x$. Next subtract $13 - 6$ on the left side to get: $7 = 12 - 5x$. Then subtract 12 from each side to yield: $-5 = -5x$. Finally, divide 5 from each side to get: $+1 = x$.

13. **EXPLANATION: The correct answer is 59.5.**

Start by adding the values together to get the length of the whole line: $2x + 1 + 3x - 1$ gives you line AD. The line's total length is therefore $5x$, so $5x = 75$ and $x = 15$. Now you can use that to find BC, which is half of AC: $2(15) + 1 = \frac{31}{2}$. That number can be added to $3(15) - 1 = 44$, which is CD. $\frac{31}{2} + 44 = 59.5$, which can be gridded in as a decimal.

14. **EXPLANATION: The correct answer is $\frac{7}{6}$ or 1.16 or 1.17.**

Put all your fractions into common denominator—something the SAT expects you to do a lot!

$\frac{1}{3}x = \frac{2}{6}x$, so the left side simplifies to $\frac{5}{6}x - \frac{2}{6}x = \frac{3}{6}x$.

For the right side, $\frac{3}{4} = \frac{9}{12}$ and $\frac{1}{6} = \frac{2}{12}$, so you subtract $\frac{9}{12} - \frac{2}{12} = \frac{7}{12}$.

Now you can set them equal to each other: $\frac{3}{6}x = \frac{7}{12}$.

Now simplify: $\frac{3}{6}x = \frac{1}{2}x$.

Then solve $\frac{1}{2}x = \frac{7}{12}$ by multiplying each side by $\left(\frac{2}{1}\right)$: $\left(\frac{2}{1}\right)\frac{1}{2}x = \frac{7}{12}\left(\frac{2}{1}\right)$.

This simplifies into $x = \frac{7}{12}\left(\frac{2}{1}\right) = \frac{14}{12}$, so $x = \frac{14}{12}$. Reduce this, so $x = \frac{7}{6}$.

You can grid this as an improper fraction, so don't worry: the SAT will count it as correct. However, if you prefer decimals, that's fine too. Remember that you can either round the decimal correctly to fill the grid in bubbles (1.17) or put in as many places as you have without rounding (1.16). Do not try to grid as a mixed number! The SAT will interpret $1\frac{1}{6}$ as the improper fraction $\frac{11}{6}$ and mark your answer wrong.

15. **EXPLANATION: Choice A is correct.**

In order to solve, you simply need to distribute and then combine like terms: $8x + 16 - 5 = 8x + 11$.

16. **EXPLANATION: Choice D is correct.**

Solve for b first: the second equation tells us $b = 24$. Now substitute that into the first equation to learn that $a = 28$. $a + b = 52$. If you got one of the other answers, you likely did just part of the problem. These are common traps, so watch out for them!

17. **EXPLANATION: Choice B is correct.**

Don't be confused by the fact that we are dealing with functions here: the concept is the same as all the other problems. Isolate $f(x)$ by adding $g(x)$ to both sides of the equation.

18. **EXPLANATION: Choice A is correct.**

Substitute 70 in for $f(x)$ to get $70 = 20 + 2.5y$. Then subtract 20 from both sides to get $50 = 2.5y$. Next, divide each side of the equation by 2.5 to get $y = 20$.

19. **EXPLANATION: Choice C is correct.**

Just isolate the variable you're looking for: to get h alone, you need to divide both sides of the equation by $9w$, which gives you answer choice C.

20. **EXPLANATION: Choice D is correct.**

This question is a bit tricky because it has two parts. Start by solving for h: $3h + 15 = 6$, so $3h = 6 - 15$. Therefore, $3h = -9$ and $h = -3$. Then, add $2(-3)$ to 10 to get 4. If you got one of the other answer choices, you may have made a sign error or just done one part of the problem.

21. **EXPLANATION: Choice D is correct.**

This has a fraction in it, so multiply both sides by 4 to remove the fraction to get $x - 2 = 4n$. Substitute the n for 5 and you'll get $x - 2 = 20$. Then, you can solve to find $x = 22$ by adding 2 on the left side and adding it to the right side. If you stopped here, you would get answer choice C, 22, but you need to go one step further to find $x + 5$. Since we know that $x = 22$, you need to add 5 to get 27. Don't forget that as a backup plan, you can try plugging in the answers one by one to the equation to find one that works for you—that'll work for many algebra questions!

CHAPTER 5

EXPONENTS AND RADICALS

The secret to solving this type of problem is to know the power and root algebraic rules. The best way to do this is to apply the rules to problems. This section gives you the practice you need to learn these algebraic techniques. Begin with the mathematics review and then complete and correct the practice problems. There are 2 Solved SAT Problems and 10 Practice SAT Questions with answer explanations.

EXPONENTS

A whole number exponent shows how man times to repeated multiply the base. $_{\text{Base}}5^{4\ \text{Exponent}} = 5 \cdot 5 \cdot 5 \cdot 5$.

Calculator $5^4 = 5 \wedge 4$

A radical is the opposite of an exponent.

This radical sign ($\sqrt{}$ —square root means "what number multiplied by itself equals the number inside."

Square 3 to get 9: $3^2 = 9$. Find the square root of 9 to get 3. $\sqrt{9} = 3$.

Here are some more examples.

$2^2 = 4, \sqrt{4} = 2.$ $\qquad 4^2 = 16, \sqrt{16} = 4.$

A number outside the radical sign like this shows the root is not 2.

$^{\text{Root}} \sqrt[3]{8}$

$2^3 = 2 \times 2 \times 2 = 8$, So $\sqrt[3]{8} = 2$

$3^4 = 3 \times 3 \times 3 \times 3 = 81$, So $\sqrt[3]{81} = 3$

Exponents can be fractions. The denominator of the exponent becomes the root. Here are some examples.

$2^{\frac{1}{2}} = \sqrt{2}, \quad 2^{\frac{1}{3}} = \sqrt[3]{2}$

$2^{\frac{6}{3}} = \sqrt[3]{2^6} = \sqrt[3]{64} = 4$

$2^{\frac{4}{3}} = \sqrt[3]{2^4} = \sqrt[3]{16} = \sqrt[3]{8 \times 2} = \sqrt[3]{8} \times \sqrt[3]{2} = 2\sqrt[3]{2}$

Below are some basic rules for exponents and radicals.

$x^0 = 1$ $\qquad\qquad x^a \cdot x^b = x^{a+b}$ $\qquad\qquad \dfrac{x^a}{x^b} = x^{a-b}$

$(x^a)^b = x^{a \times b}$ $\qquad\qquad x^{-a} = \dfrac{1}{x^a}$

$(x \cdot y)^a = x^a \cdot y^a$ $\qquad\qquad x^{\frac{a}{b}} = \left(\sqrt[b]{x}\right)^a = \sqrt[b]{x^a}$

$\sqrt{a \times b} = \sqrt{a} \times \sqrt{b}$ $\qquad\qquad \sqrt{\dfrac{a}{b}} = \dfrac{\sqrt{a}}{\sqrt{b}}$

These examples use the rules of exponents. Not every step is shown.

Example 1.

$x^2 \cdot x^b = x^9, b =$

$\quad x^2 \cdot x^b = x^{2+b} = x^9 \Rightarrow 2 + b = 9 \Rightarrow b = 7.$

Example 2.

$(2^3)^b = 64, b =$

$\quad (2^3)^b = 2^{3 \times b} = 64 = 2^6 \Rightarrow 3 \times b = 6 \Rightarrow b = 2.$

Example 3.

$\dfrac{a^3 b^7 c^{11}}{a^{10} b^8 c^4} =$

$\dfrac{a^3 b^7 c^{11}}{a^{10} b^8 c^6} = a^{-7} b^{-1} c^5 = \dfrac{c^5}{a^7 b}.$

Example 4.

If $3^{-a} = 81$ what is the value of a.

We know $3^4 = 81$ so a must equal -4.

Example 5.

$x^{\frac{3}{2}} = 8, x =$

$x^{\frac{3}{2}} = 8 \Rightarrow \left(x^{\frac{3}{2}}\right)^{\frac{2}{3}} = 8^{\frac{2}{3}} \Rightarrow x^{\frac{3}{2} \times \frac{2}{3}} = 8^{\frac{2}{3}}$

$\quad x = 8^{\frac{2}{3}} = \left(\sqrt[3]{8}\right)^2 = 2^2 = 4 \Rightarrow x = 4.$

Simplify radicals.

Example 6.

$4\sqrt{50} - 3\sqrt{32} =$

$4\sqrt{50} - 3\sqrt{32} = 4\sqrt{25 \times 2} - 3\sqrt{16 \times 2} = 4 \times \sqrt{25} \times \sqrt{2} - 3 \times \sqrt{16} \times \sqrt{2} =$

$\quad 4 \times 5 \times \sqrt{2} - 3 \times 4 \times \sqrt{2} = 20\sqrt{2} - 12\sqrt{2} = 8\sqrt{2}.$

Example 7.

$$\frac{9\sqrt{24}}{2\sqrt{27}} =$$

$$\frac{9\sqrt{24}}{2\sqrt{27}} = \frac{9\sqrt{4 \times 6}}{2\sqrt{9 \times 3}} = \frac{9 \times 2\sqrt{6}}{2 \times 3\sqrt{3}} = \frac{18\sqrt{6}}{6\sqrt{3}} = 3\sqrt{2}.$$

Example 8.

$$7\sqrt{x} + 14 = 42, x =$$

$$7\sqrt{x} + 14 = 42 \Rightarrow 7\sqrt{x} = 28 \Rightarrow \sqrt{x} = 4 \Rightarrow x = 16.$$

Practice Questions

1. $\dfrac{x^a}{x^{15}} = \dfrac{1}{x^{-5}}, a =$

2. $(a^2)^4 = 6{,}561, a =$

3. $\dfrac{a^{14}b^8c^6}{a^5b^{13}c^{15}} =$

4. $27^{\frac{5}{3}} = x^5, x =$

5. $7\sqrt{45} + 5\sqrt{180} =$

6. $\dfrac{3\sqrt{60}}{2\sqrt{20}} =$

7. $8\sqrt{x} - 37 = -21, x =$

Practice Answers

1. $a = 20.$
2. $a = 3.$
3. $a^9 b^{-5} c^{-9} = \dfrac{a^9}{b^5 c^9}.$
4. $x = 3.$
5. $51\sqrt{5}.$
6. $\dfrac{3\sqrt{3}}{2}.$
7. $x = 4.$

SOLVED SAT PROBLEMS

1. $16y = 2^{4+x}$, $y =$

   A. 2^x
   B. $2x$
   C. 2
   D. 4

   EXPLANATION: The correct answer is A.

   Use the rules of exponents.

   $16y = 2^{4+x} = 2^4 \cdot 2^x = 16 \cdot 2^x$

   That means $16y = 16 \cdot 2^x \Rightarrow y = 2^x$.

2. $\sqrt{-a} - 1 = 7$, $a =$

   A. -64
   B. 64
   C. 49
   D. 36

   EXPLANATION: The correct answer is A.

   Use the rules of exponents.

   $\sqrt{-a} - 1 = 7 \Rightarrow \sqrt{-a} = 8$

   We know $\sqrt{64} = 8$.

   That means $-a = 64$ and $a = -64$.

EXPONENTS AND RADICALS
PRACTICE SAT QUESTIONS

ANSWER SHEET

Choose the correct answer.

1. Ⓐ Ⓑ Ⓒ Ⓓ
2. Ⓐ Ⓑ Ⓒ Ⓓ
3. Ⓐ Ⓑ Ⓒ Ⓓ
4. Ⓐ Ⓑ Ⓒ Ⓓ
5. Ⓐ Ⓑ Ⓒ Ⓓ
6. Ⓐ Ⓑ Ⓒ Ⓓ
7. Ⓐ Ⓑ Ⓒ Ⓓ
8. Ⓐ Ⓑ Ⓒ Ⓓ
9. Ⓐ Ⓑ Ⓒ Ⓓ
10. Ⓐ Ⓑ Ⓒ Ⓓ

PRACTICE SAT QUESTIONS

1. $\sqrt{x-5}+6=11, x=$
    A. 5
    B. 20
    C. 25
    D. 30

2. $3^{\frac{x}{2}}=27, x=$
    A. 1.5
    B. 3
    C. 4.5
    D. 6

3. $9^a \cdot 3^c = 177,147. \quad 2a+c=$
    A. 8
    B. 9
    C. 10
    D. 11

4. Which of the following is equal to $b^{\frac{4}{5}}$ for all values of b?
    A. $\sqrt{b^4}$
    B. $\sqrt{b^5}$
    C. $\sqrt[4]{b^5}$
    D. $\sqrt[5]{b^4}$

5. If $\sqrt{p}+\sqrt{16}=\sqrt{81}$, what is the value of p?
    A. 2
    B. 5
    C. 9
    D. 25

6. If $5x-y=7$, what is the value of $\dfrac{32^x}{2^y}$?
    A. 2^7
    B. 4^7
    C. 8^3
    D. The answer cannot be determined from the given information.

7. If $c^{\frac{d}{5}}=25$ for positive integers c and d, what is one possible value of d?
    A. 0
    B. 10
    C. 27
    D. 40

8. $4x^8 + 16x^4y + 16y^2$

    Which of the following is equivalent to the expression shown above?
    A. $2(8x^4y + 4y^4)^2$
    B. $(4x + 4y)^8$
    C. $(2x^2 + 4y^2)^4$
    D. $4(x^4 + 2y)(x^4 + 2y)$

9. $25c^8 + 40c^4d^4 + 16d^8$

    Which of the following is equivalent to the expression shown above?
    A. $(5c^4 + 4d^4)^2$
    B. $(5c + 4d)^8$
    C. $(25c^4 + 16d^4)^2$
    D. $(5c^4 + 4d^4)^4$

10. $\sqrt{4n^2 + 13} - y = 0$

    If $n > 0$ and $y = 7$ in the equation above, what is the value of n?
    A. 2
    B. 3
    C. 4
    D. 5

EXPLAINED ANSWERS

1. **EXPLANATION: Choice D is correct.**

   Simplify the equation $\sqrt{x-5}+6=11$ by subtracting 6 to get $\sqrt{x-5}=5$; square both sides $(\sqrt{x-5})^2=(5)^2$; $x-5=25$; solve for x; $x=30$. Alternatively, back-solve by plugging the answers back into the equation.

2. **EXPLANATION: Choice D is correct.**

   First, find the power of 3 that equals 27; $3^3=27$, so $3^{\frac{x}{2}}=3^3$. Next, let $\frac{x}{2}=3$. Multiply both sides by 2 to get $x=6$.

3. **EXPLANATION: Choice D is correct.**

   The key here is to see that you can rewrite 9 as 3^2. So $(9^a)(3^c)=(3^2)^a\,(3^c)=(3^{2a})\,(3^c)=(3^{2a+c})=177{,}147$. Use the guess-and-check method on your calculator to find the power of 3 that will result in 177,147. $3^{11}=177{,}147$, so $3^{2a+c}=177{,}147$. Therefore, $2a+c=11$.

4. **EXPLANATION: Choice D is correct.**

   It's important to know your exponent rules: with a fractional exponent, the denominator becomes the root, outside the radical, and the top number becomes the exponent, inside the radical.

   Root$\rightarrow\sqrt[5]{b^4}\leftarrow$Exponent

5. **EXPLANATION: Choice D is correct.**

   The math on this question isn't hard, but be sure you work methodically. The square root of 16 is 4, and that of 81 is 9, so the square root of p needs to be 5. Therefore, $p=25$. Notice that 5 and 9, which are partial answers to various parts of the problem, are answer choices there to trap you. Don't fall for these tricks!

6. **EXPLANATION: Choice A is correct.**

   First of all, ignore answer choice D. This is rarely the answer on a math question, especially on a more difficult one like this; they're hoping that you won't know what to do and will select that answer. For this question, start by converting your exponents into the same base. In this case, 2. 2 to the 5th power is 32, so that becomes $\frac{2^{5x}}{2^y}$. According to the rules of exponents, when one divides exponents, one must subtract, so the equation becomes 2^{5x-y}. Luckily, the question gave us an additional piece of information, that $5x-y=7$. So the answer is 2^7.

7. **EXPLANATION: Choice B is correct.**

   Don't be afraid of the fact that there are two variables here. Make your life easier by picking something easy for c, like 5 (a number that is going to easily get to 25 when you raise it to a power). The power you need to raise 5 to get an answer of 25 is 2, so $\frac{d}{5}$ must be equal to 2. At that point, solve for d or use the answer choices and plug in to find the correct answer.

   Alternatively, plug in your answer choices for d. Check to see if an answer choice works by seeing if the value for c would be an integer. The only answer choice that gives an integer for c is $d=10$.

8. **EXPLANATION: Choice D is correct.**

   Each coefficient is divisible by 4. Factor out the 4: $(4x^8+16x^4y+16y^2)=4(x^8+4x^4y+4y^2)$. D is the only answer that looks like that. We could stop here. But let's check. Try Answer D. Use FOIL to multiply. Remember: to multiply two exponents with the same base, add the exponents and keep the base, $4\,(x^4+2y)\,(x^4+2y)=4(x^8+4x^4y+4y^2)$. Now multiply by 4. $4(x^8+4x^4y+4y^2)=4x^8+16x^4y+16y^2$. That's it. D is the correct answer.

9. **EXPLANATION: Choice A is correct.**

You could factor this trinomial, but let's work from the answers. Look at the first term and the power outside the parentheses and the outside power in each answer choice. The coefficient of the first term in the answer choice, raised to the outside power, must equal 25.

A. $(\underline{5c^4} + 4d^4)^2$ $\Rightarrow (5c^4)^2 = (5^2)(c^4)^2 = 5^2 = 25$. Could be correct.
B. $(\underline{5c} + 4d)^8$ $\Rightarrow (5c)8 = (5^8)(c)^8 = 5^8 \neq 25$. Cannot be correct.
C. $(\underline{25c^4} + 16d^4)^2 \Rightarrow (25c^4)^2 = (25^2)(c^4)^2 = 25^2 \neq 25$. Cannot be correct.
D. $(\underline{5c^4} + 4d^4)^4$ $\Rightarrow (5c^4)^2 = (5^4)(c^4)^4 = 5^4 \neq 25$. Cannot be correct.

Choice A is the only correct answer. Don't forget to be strategic about how you use your answer choices: as here, it can save you time and effort!

This trinomial is easy to factor. Here's a quick explanation.

$$25c^8 + 40c^4d^4 + 16d^8$$

$(\underline{5c^4}\ \ )(\underline{5c^4}\ \ )$ Factor the first term of the trinomial.

$(5c^4 + \underline{4d^4})(5c^4 + \underline{4d^4})$ Factor the last term so that sum of the numerical factors multiplied by 5 = 40. That's $4d^4 \times 4d^4$.

$(5c^4 + 4d^4)^2$ Write with an exponent.

10. **EXPLANATION: Choice B is correct.**

Substitute 7 for y and then add it to the other side of the equation to get $\sqrt{4n^2 + 13} = 7$. Next, square both sides to get rid of the radical: $4n^2 + 13 = 49$. Now simplify the equation, subtracting 13 from both sides to get $4n^2 = 36$, and divide both sides by 4 to get $n^2 = 9$. Therefore, $n = 3$.

CHAPTER 6

SYSTEMS OF EQUATIONS

When solving systems of equations, typically you are dealing with two variables; call the variables x and y. The goal is to use techniques to isolate one of the variables, for example x, solve for x, and then if necessary, substitute the value for x into one of the equations to solve for y. The SAT systems of equations problems might have the equations already set up or it may be a word problem that requires you to set up the equations. Begin with the mathematics review and then complete and correct the practice problems. There are 2 Solved SAT Problems and 23 Practice SAT Questions with answer explanations.

SOLVING SYSTEMS OF EQUATIONS

Systems of equations are sometimes called simultaneous equations. The solution to a system of equations may also be the point (x, y) where two lines or two line segments cross.

Example 1.

Find the ordered pair (x, y) that satisfies the system of equations shown below.

$$2x + 4y = 10$$
$$x - 2y = -3.$$

Method 1: (Substitution)

Solve for x in the second equation, and substitute for x into the first equation:

$$x - 2y = -3 \implies x = 2y - 3$$

Substitute $(2y - 3)$ for x in the first equation.

$$2(2y - 3) + 4y = 10$$

Solve for y.

$$2(2y - 3) + 4y = 10 \implies 4y - 6 + 4y = 10$$
$$8y - 6 = 10 \implies 8y = 16 \implies y = 2$$

Because $x = (2y - 3)$, and $y = 2$, $x = 2(2) - 3 = 4 - 3 = 1$. The solution to this system of equations is $x = 1$ and $y = 2$. Check the solution. Substitute the values for x and y into either original equation.

Method 2: (Elimination)

Multiply the second equation by 2 to make elimination easier.

$$2x + 4y = 10$$
$$2(x - 2y) = 2(-3)$$

Add the two equations to eliminate the y and solve for x.

$$2x + 4y = 10$$
$$\underline{2x - 4y = -6}$$
$$4x + 0y = 4 \;\Rightarrow\; 4x = 4 \;\Rightarrow\; x = 1$$

Substitute the value of x into either of the original equations to find the value of y.

$$2x + 4y = 10 \;\Rightarrow\; 2(1) + 4y = 10 \;\Rightarrow\; 2 + 4y = 10$$
$$4y = 8 \;\Rightarrow\; y = 2$$

OR

$$x - 2y = -3 \;\Rightarrow\; 1 - 2y = -3$$
$$-2y = -4 \;\Rightarrow\; y = 2.$$

SYSTEMS OF EQUATIONS

Example 2.

Find the ordered pair (x, y) that satisfies the system of equations shown below.

$$y = x + 3$$
$$x + 2y = 15$$

Based on the setup of the system, substitute $y = x + 3$ into the second equation.

$$x + 2(x + 3) = 15$$

Solve for x: $x + 2x + 6 = 15 \;\Rightarrow\; 3x = 9 \;\Rightarrow\; x = 3$
To solve for y, substitute $x = 3$ back into first equation: $y = x + 3 = 3 + 3 = 6$
So the ordered pair is $(3, 6)$.

Example 3.

If (x, y) is a solution that satisfies the system of equations shown below, what is the value of y?

$$-2x + 5y = 8$$
$$4x - 3y = -2$$

Based on the setup of the system, eliminate the x.
Multiply the first equation by 2: $2(-2x + 5y) = 2(8) \;\Rightarrow\; -4x + 10y = 16$
The new system is:

$$-4x + 10y = 16$$
$$4x - 3y = -2$$

Add the two equations to eliminate x and solve for y:

$$-4x + 10y = 16$$
$$+ \quad 4x - 3y = -2$$
$$\overline{ 0x + 7y = 14} \quad \Rightarrow \quad 7y = 14 \quad \Rightarrow \quad y = 2$$

The question just asks for the value of y. We're done.

Example 4.

Last month Liam sold 9 more cars than Colin. Together they sold a total of 45 cars. How many cars did Liam sell last month?

Let L = Number of cars Liam sold

Let C = Number of cars Colin sold

Set up a system to describe the situation:

$$L = C + 9$$
$$L + C = 45$$

Substitute $L = C + 9$ into the second equation.

$$(C + 9) + C = 45$$

Solve for C: $2C + 9 = 45 \quad \Rightarrow \quad 2C = 36 \quad \Rightarrow \quad C = 18$

To solve for L, substitute $C = 18$ back into the first equation:

$$L = C + 9 = 18 + 9 = 27$$

Liam sold 27 cars.

Alternatively, you can also solve for $C = L - 9$ and then substitute into the second equation to solve for L directly.

Example 5.

The toll for a bridge is \$12.50 per vehicle unless 3 or more people in the vehicle. That is a carpool, which discounts the toll to \$6.50 per vehicle. If the total in tolls collected on a given day was \$9,125 and a total of 850 cars traveled over the bridge, how many carpool vehicles traveled over the bridge?

Let x = Number of carpool vehicles

Let y = Number of regular vehicles

Set up a system to describe the situation:

$$x + y = 850$$
$$6.5x + 12.5y = 9,125$$

Either substitution or elimination will work. This example uses elimination. Eliminate either x or y. We eliminate the y.

Multiply the first equation by 12.5:

$$12.5(x + y) = 12.5(850) \quad \Rightarrow \quad 12.5x + 12.5y = 10,625$$

The new system is:

$$12.5x + 12.5y = 10,625$$
$$6.5x + 12.5y = 9,125$$

Subtract the two equations to eliminate y and solve for x:

$$12.5x + 12.5y = 10,625$$
$$-\quad 6.5x + 12.5y = 9,125$$
$$6x + 0y = 1,500 \implies 6x = 1,500 \implies x = 250$$

There are 250 carpool vehicles.

Practice Questions

1. Solve the system of equations for x and y.

   $$-3x - 2y = 8$$
   $$5x + y = 3.$$

2. Find the ordered pair (x, y) that satisfies the system of equations shown below.

   $$\frac{8}{x} = 4$$
   $$5x - 2y = 24.$$

3. Brendan and Lindsey each work at the local grocery store. Last week Brendan and Lindsey worked a total of 40 hours and Lindsey worked 4 more hours than Brendan. How many hours did Brendan work and how many hours did Lindsey work?

4. Tickets for a concert went on sale at two different prices, $25 for an Adult ticket and $15 for a Child ticket. A total of 1,000 tickets were sold at a total sales cost of $18,000. How many Adult tickets were sold and how many Child tickets were sold?

Practice Answers

1. $x = 2$ and $y = -7$.
2. $x = 2$ and $y = -7$. (Yes! Questions 1 and 2 have the same answer.)
3. Brendan worked 18 hours and Lindsey worked 22 hours.
4. 300 Adult tickets were sold and 700 Child tickets were sold.

SOLVED SAT PROBLEMS

1. If $2ab - 6a = -6$ and $\dfrac{6}{a} = 2$, $b =$

   A. 1
   B. 2
   C. 3
   D. 4

   EXPLANATION: Choice B is correct.

   Solve for a first. Since $\dfrac{6}{a} = 2$, $a = 3$.

   Substitute $a = 3$ into $2ab - 6a = -6$: $2(3)b + 6(3) = -6$.

   Simplify: $6b - 18 = -6$.

   Add 18 to both sides of equation: $6b = 12$.

   Divide both sides of equation by 6: $b = 2$.

2. In a game, scoring with a blue ball earns a team 2 points and scoring with a red ball earns a team 3 points. Team 1 scored a total of 32 points and Team 2 scored a total of 28 points. If both teams scored the same number of red balls, how many more blue balls did Team 1 score than Team 2?

A. 1
B. 2
C. 3
D. 4
E. 5

EXPLANATION: Choice B is correct.

Define the variables.

R = Number of red balls scored by each team

B_1 = Number of blue balls for Team 1

B_2 = Number of blue balls for Team 2

Write simultaneous equations to describe the relationship.

$$2B_1 + 3R = 32$$
$$2B_2 + 3R = 28$$

Subtract the equations.

$$2B_1 + 3R = 32$$
$$\underline{-2B_2 + 3R = 28}$$
$$2B_1 - 2B_2 = 4$$

Solve for $B_1 - B_2 \Rightarrow 2B_1 - 2B_2 = 4 \Rightarrow 2(B_1 - B_2) = 4$

$$B_1 - B_2 = 2.$$

Team 1 scored two more blue balls than Team 2.

SYSTEMS OF EQUATIONS
PRACTICE SAT QUESTIONS

ANSWER SHEET

Choose the correct answer.
If no choices are given, grid the answers in the section at the bottom of the page.

1. (A) (B) (C) (D)
2. (A) (B) (C) (D)
3. (A) (B) (C) (D)
4. (A) (B) (C) (D)
5. (A) (B) (C) (D)
6. (A) (B) (C) (D)
7. GRID
8. (A) (B) (C) (D)
9. (A) (B) (C) (D)
10. (A) (B) (C) (D)

11. GRID
12. GRID
13. GRID
14. GRID
15. (A) (B) (C) (D)
16. (A) (B) (C) (D)
17. (A) (B) (C) (D)
18. (A) (B) (C) (D)
19. (A) (B) (C) (D)
20. (A) (B) (C) (D)

21. GRID
22. (A) (B) (C) (D)
23. (A) (B) (C) (D)

Use the answer spaces in the grids below if the question requires a grid-in response.

Student-Produced Responses — ONLY ANSWERS ENTERED IN THE CIRCLES IN EACH GRID WILL BE SCORED. YOU WILL NOT RECEIVE CREDIT FOR ANYTHING WRITTEN IN THE BOXES ABOVE THE CIRCLES.

7.

11.

12.

13.

14.

21.

PRACTICE SAT QUESTIONS

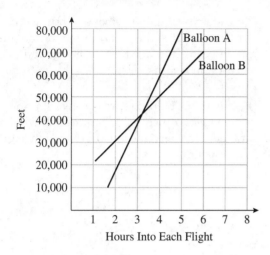

Feet — Hours Into Each Flight

1. Two hot-air balloons are climbing up from the desert floor on a hot day at regular rates. As shown on the graph above, Balloon A reached a height of 80,000 feet. Balloon B reached a height of 70,000 feet. About how many hours into the flights were the balloons at the same height?

   A. 3
   B. 4
   C. 30,000
   D. 40,000

Questions 2 and 3 refer to this scenario.

A bulb manufacturer creates red bulbs on two production lines. The equations below show the number of red bulbs (R_1 and R_2) produced on each line, where $t > 0$ is the time in minutes.

$$R(1) = \frac{4}{5}t + 20,$$
$$R(2) = 48 + \frac{3}{5}t$$

2. By how much does the number of bulbs produced in Line 1 decrease if t decreases by 10 minutes?

   A. 2
   B. 4
   C. 8
   D. 10

3. To the nearest minute, at what time (t) will the production in Line 1 equal the production in Line 2?

   A. 6
   B. 68
   C. 140
   D. 340

4. $x = y - 1$ and $\dfrac{x}{2} + 3y = 10$

   Which ordered pair (x, y) satisfies the system of equations shown above?

   A. (−2, 3)
   B. (2, −3)
   C. (3, 2)
   D. (2, 3)

5. The trash haulers charge $7 for each container of trash and $1 for each container of recycling. There were a total of 9 containers of trash and recycling. The total charge before taxes for this entire collection was $45. How many containers of trash were in this collection?

   A. 4.5
   B. 6
   C. 9
   D. 54

6. The shipping company packs items two in a box or six in a box. On this delivery, there are 58 items to be shipped and there are a total of 13 shipping boxes. How many of the boxes contain two items?

   A. 0
   B. 2
   C. 5
   D. 6

7. The delicatessen charges $6.95 a pound for turkey breast and $4.95 a pound for chicken breast. A customer buys 8 pounds combined of the two meats and does not buy anything else at the deli. She is charged a price of $51.60. How many pounds of turkey breast did the customer buy?

8. A sports trading card company produces two sets of cards. One type with 40 cards is the modern (m) edition, which features information about current sports figures. The other type, with 30 cards, is the historic (h) edition, which features sports figures who are no longer active. The company prints a combined total of 100 sets of cards that ship as individual cards and take 50 containers to ship. Each container holds exactly 72 cards. Which of the following sets of equations can be used to find the number of modern sports cards printed?

A. $m = 100 - h$
   $40m + 30h = 3,600$
B. $m = 100 - h$
   $40m + 30h = 100$
C. $m = 100 - h$
   $30m + 40h = 3,600$
D. $m + h = 50$
   $40m + 30h = 3,600$

9. A theme park features two attractions that include moving cars. The Speedway lasts 15 minutes and costs \$26. City Motors lasts 12 minutes and costs \$19. In one day, the theme park sold a total of 600 tickets for these two attractions for a total amount of \$12,800. Which of the following systems of equations could be used to find the number of tickets sold for the Speedway?

A. $15S + 12C = 12,800$
   $S + C = 600$
B. $S + C = 12,800$
   $26S + 19C = 600$
C. $(26S) + (19C) = 12,800$
   $S + C = 600$
D. $(19S) + (26C) = 12,800$
   $S + C = 600$

10. $p + q = 8$
    $q \times 2 = 18$

    What is $q - p$?
    A. -1
    B. 1
    C. 8
    D. 10

11. $2x + 4y = 1,000$
    $4x + 2y = 1,100$

    Based on the system of equations above, what is the value of $6x + 6y$?

12. $1.2x - 1.2y = 0.6$
    $0.8x + 2.4y = 2.6$

    The system of equations above is graphed in the xy-plane. What is the x-coordinate of the intersection point (x, y) of the system? Round to the nearest hundredths.

13. The equations below represent two lines in the xy-plane. For which value of r are the two lines parallel? Disregard the negative sign when gridding in your answer.

$$6x - 4y = 12$$
$$rx + 7y = 28$$

14. $cx + ky = 20$
    $4x + 16y = 80$

    In the system shown, c and k are integer constants with a combined total value of 5. If the system has infinite solutions, find the value for $\dfrac{c}{k}$.

15. $\dfrac{1}{4}(3x + y) = \dfrac{42}{5}$
    $y = 3x$

    If the solution to the system of equations above is (x, y), what is the value of x?

    A. $\dfrac{7}{75}$
    B. $\dfrac{5}{28}$
    C. $\dfrac{28}{5}$
    D. $\dfrac{63}{5}$

16. Solve the system of equations below for y.
$$-4x + 3y = 25$$
$$2x + 6y = 25$$

    A. -2.2
    B. -5
    C. 2.5
    D. 5

17. Solve the system of equations below for y:

$$-2x + y = -6.2$$
$$2x + 4y = 8.2$$

A. $y = \dfrac{2}{5}$

B. $y = \dfrac{3}{5}$

C. $y = \dfrac{5}{2}$

D. $y = \dfrac{5}{3}$

18. Find the solution to this system of equations.

$$6x + 3y = -24$$
$$3y - 2x = -16$$

A. $(-1, -6)$
B. $(-6, -1)$
C. $(6, 1)$
D. $(1, 6)$

19. The solution to the system of equations below has the form (x, y). Find the value of x.

$$\frac{1}{4}y = 6$$
$$x - \frac{1}{4}y = 3$$

A. -24
B. 3
C. 9
D. 27

20. At a local supermarket, each pound of sirloin costs $1.50 a pound more than each pound of chuck. If 2 pounds of sirloin and 3 pounds of chuck cost a total of $23.00, how much does the sirloin cost per pound?

A. $2.99/pound
B. $4.00/pound
C. $5.00/pound
D. $5.50/pound

21. The Homeowners Association authorized the expenditure of $300 to $400 to repair swings in the playground. It costs $20 to replace a swing chain (x) and $30 to replace a swing seat (y). If the playground committee decides to replace 4 swing seats, what is one possible answer for how many swing chains (x) can be replaced, staying within the authorized amount?

22. The score on a practice algebra test is obtained by subtracting the number of incorrect answers from twice the number of correct answers. If a student answered 50 questions and got a score of 70, how many questions did the student answer correctly?

A. 20
B. 30
C. 40
D. 60

23. $S = 85.50 + 0.15x$

$C = 75.50 + 0.25x$

In the equations above, C and S represent the cost of car rentals and SUV rentals per day, for x days after Memorial Day last summer. What was the cost of the rentals when the costs were equal?

A. $100
B. $100.50
C. $175.50
D. $185.50

EXPLAINED ANSWERS

1. **EXPLANATION: Choice A is correct.**

   Look for where the lines representing the two balloons cross each other: somewhere between 3 and 4 hours (which you can see along the x-axis of the graph). It's closer to 3, so that's the answer. If you picked one of the larger numbers in answer choices C or D, you looked at the wrong axis. Be sure to watch your variables.

2. **EXPLANATION: Choice C is correct.**

   Time (t) decreases by 10, so use ($t - 10$). This is Line 1, so plug that into the first equation: $\frac{4}{5}(t - 10) + 20$. Simplify to find $\frac{4}{5}t - 8 + 20 = \frac{4}{5}t + 12$. Notice that $20 - 12 = 8$, so that's a decrease by 8 on the right-hand side of the equation. So, the left-hand side, which shows the number of bulbs produced, will decrease by 8 as well.

3. **EXPLANATION: Choice C is correct.**

   To solve this type of problem, which is common on the SAT, set the two equations equal to each other and simplify.

   $\frac{4}{5}t + 20 = 48 + \frac{3}{5}t.$

   $\frac{4}{5}t = 28 + \frac{3}{5}t$ Subtract 20 from both sides of the equation.

   $\frac{1}{5}t = 28$ Subtract $\frac{3}{5}t$ from both sides of the equation.

   $t = 140$ Multiply both sides of the equation by 5.

   Each line produces the same number of bulbs at 140 minutes.

4. **EXPLANATION: Choice D is correct.**

   This method is quickest. Use the first equation $x = y - 1$. Notice that x is one less than y.

   That is only for answer choice D, the correct answer.

   If you did not see that, there are other successful approaches.

   For this problem, the next most successful method is substitution.

   Use the equations $x = y - 1$ and $\frac{x}{2} + 3y = 10$.

   Substitute $y - 1$ for x in the second equation. $y - \frac{1}{2} + 3y = 10$

   Multiply both sides of the equation by 2. $y - 1 + 6y = 20$

   Simplify $7y - 1 = 20 \Rightarrow 7y = 21 \Rightarrow y = 3$.

   Substitute 3 for y in $x = y - 1 \Rightarrow x = 3 - 1 \Rightarrow x = 2$.

   $x = 2, y = 3$. That's (2, 3), answer choice D.

   Most systems of equations on the SAT can also be solved by the elimination method—add or subtract the equations and solve. It's just harder for this problem because the first equation needs to be rewritten as $x - y = -1$. You can also work back from the answers, substituting the values for x and y into the equations. Use what works best for you.

5. **EXPLANATION: Choice B is correct.**

   Create two equations from the facts in the problem. To eliminate the r variable, multiply by –1, as shown below, where t represents trash and r represents recycling:

   ORIGINAL EQUATIONS:

   $7t + r = 45$

   $t + r = 9$

   $-1(t + r = 9)$ means that your two new equations will be:

   $7t + r = 45$

   $-t - r = -9$

   Add the equations: $6t = 36$

   $t = 6$ containers of trash

   If you picked answer choice A, you may have multiplied $(-1)\, t$ and wrote "t" instead of "$-t$," or you may have added $7t + (-t)$ and wrote $8t$ instead of $6t$. If you chose answer choice C, you may have added $45 + (-9)$ and wrote 54 instead of 36.

6. **EXPLANATION: Choice C is correct.**

   This one may be easier to solve by substitution. Create two separate equations:

   $2x + 6y = 58$

   $x + y = 13$

   Solve the bottom equation for one of the variables: $\quad y = 13 - x$

   Substitute this value for y in the first equation: $\quad 2x + 6y = 58$

   $\quad 2x + 6\,(13 - x) = 58$

   Distribute the variable: $\quad 2x + 78 - 6x = 58$

   Simplify and solve for x: $\quad -4x = -20$

   Divide 4 from each side and solve for x: $\quad x = 5$

   Therefore, the answer is five boxes with two items.

   This problem could also be solved by manipulating the equation pairs and eliminating one of the variables through addition of the two equations.

7. **EXPLANATION: The correct answer is 6.**

   Create two equations. $\quad 6.95t + 4.95c = 51.60$

   $\quad t + c = 8$

   Use the substitution or the elimination method to solve. Here's a demonstration of how you could use substitution: $c = 8 - t$; $6.95(t) + 4.95(8 - t) = 51.60$; multiply $(6.95t + 39.60 - 4.95t = 51.60)$; combine like terms and subtract 39.60 from both sides. Therefore, $2t = 12$; $t = 6$ pounds.

8. **EXPLANATION: Choice A is correct.**

   The first equation in answer choice A shows the correct facts from the problem. That is, $m + h = 100$ sets, so $m = 100 - h$. On the left side of the second equation for this choice, we see the expression for the total number of modern and historic cards $(40m + 30h)$. On the right side, find 3,600, the total number of modern and historic cards. Multiply the number of containers (50) by the number of cards in each container (72). $50 \times 72 = 3,600$.

9. **EXPLANATION: Choice C is correct.**

   Only answer choice C shows two correct equations. The number of Speedway tickets (S) + the number of City tickets (C) is 600. That's $S + C = 600$. That equation is shown in answer choices A, C, and D. The correct equation, $26S + 19C = 12,800$, for the total cost of both attractions is only shown only in C. The number of minutes each attraction lasts is extra information and not needed to solve the problem.

10. **EXPLANATION: Choice D is correct.**

Use one equation at a time:
$q \times 2 = 18, q = 9$
$p + 9 = 8$
$p = -1$
$q - p = 9 - (-1) = 10.$

11. **EXPLANATION: The correct answer is 2,100.**

Add together the two equations, and you get $6x + 6y$ on the left, so the number on the right should also consist of adding the numbers together ($1,000 + 1,100 = 2,100$).

12. **EXPLANATION: The correct answer is 1.18 or 1.19.**

Use elimination to solve this system. Multiplying the top equation by 2 cancels the y values, so you get:
$2.4x - 2.4y = 1.2$
$0.8x + 2.4y = 2.6$
Now, add the two equations together: $3.2x = 3.8$
Divide both sides of the equation by 3.2 to get: $x = 1.1875$, which can be gridded in as 1.18 or 1.19.

13. **EXPLANATION: The correct answer is 10.5 (grid in 10.5).**

Parallel lines have the same slope and do not intersect. So we are looking for the value of y when the slopes are equal. Start with the equations in $y = mx + b$ form.

$$y = \frac{6}{4}x - 3$$

$$y = \frac{-r}{7}x + 4$$

Now that it's clear that the slopes are equal and $\frac{6}{4} = \frac{-r}{7}$, solve using cross-multiplication. $6(7) = 4(-r)$, $42 = -4r$ and $r = -10.5$. Remember to disregard the negative sign when you grid in the answer.

14. **EXPLANATION: The correct answer is $\frac{1}{4}$.**

Since the system above has infinitely many solutions, the equations represent the same line, and the equations must be equal.

Divide by 4 to simplify the bottom equation: $4x + 16y = 80 \Rightarrow x + 4y = 20$, which is the same as $cx + ky = 20$.
So $c = 1$ and $k = 4$. We know from the problem that $c + k = 5$ and $\frac{c}{k} = \frac{1}{4}$.

15. **EXPLANATION: Choice C is correct.**

Use substitution to solve the first equation, plugging in $3x$ for y. $\frac{1}{4}(3x + 3x) = \frac{42}{5}$; simplify: $\frac{1}{4}(6x) = \frac{42}{5}$.
Then, $\frac{3}{2}x = \frac{42}{5}$, and $\left(\frac{42}{5}\right)\left(\frac{2}{3}\right) = \frac{28}{5}$.

16. **EXPLANATION: Choice D is correct.**

This problem could be solved by either the substitution or the elimination method. Having the x variable with opposite signs already makes the elimination method an easy choice. First, multiply the second equation through by 2, which makes it $4x + 12y = 50$. Add this to the original first equation to eliminate the x variables. This leaves $15y = 75$, so $y = 5$.

17. **EXPLANATION: Choice A is correct.**

Like the last question, this one is best solved by elimination. Add both sides of the equations to eliminate the x variables. This gives you $5y = 2$. Divide by 5 to get $y = \dfrac{2}{5}$.

$$-2x + \ y = -6.2$$
$$\underline{\ 2x + 4y = \ \ 8.2\ }$$
$$5y = \ \ 2$$

Divide by 5 to solve: $\quad y = \dfrac{2}{5}$.

18. **EXPLANATION: Choice A is correct.**

To begin, rewrite the bottom equation, with an x term followed by a y term. This makes it a lot easier to use elimination to solve the problem. Multiply the second equation by 3 to get $-6x + 9y = -48$. Add the two equations together and solve for y to get $12y = -72$, or $y = -6$ as follows:

$$6x + 3y = -24 \ \Rightarrow \ 6x + 3y = -24 \quad \text{Leave unchanged.}$$
$$-2x + 3y = -16 \ \Rightarrow \ \underline{-6x + 9y = -48} \quad \text{Multiply each side by 3.}$$
$$12y = -72 \quad \text{Add equations.}$$
$$y = \ -6 \quad \text{Solve for } y.$$

Save time by looking at the answer choices. Only answer choice A has a y value of -6.

If you don't check the answer choices, you can substitute (-6) in the first equation and solve for x. $6x + 3(-6) = -24$; simplify: $6x - 18 = -24$; add (18) to both sides $(6x = -6)$. Divide by 6: $(x = -1)$. The solution is $(-1, -6)$.

19. **EXPLANATION: Choice C is correct.**

Use the substitution method. Replace $\dfrac{1}{4}y$ with 6 and then solve the second equation. $x - 6 = 3$; add 6 to both sides to solve for x. $x = 9$.

20. **EXPLANATION: Choice D is correct.**

To find the cost of the sirloin, write and solve a system of two equations. Let x equal the cost of the sirloin per pound and y equal the cost of the chuck per pound. Since the total cost of the beef was $23.00, the first equation is $2x + 3y = \$23.00$. Since the sirloin cost $1.50 more per pound than the chuck, the equation $x = y + \$1.50$ must also be true from the given facts. Use the substitution method to solve this system of equations. $2(y + \$1.50) + 3y = \$23.00$. Distribute and simplify the equation. $2y + 3 + 3y = 23.00$, so $5y + 3 = 23.00$. Subtract 3 from both sides of the equation: $5y = 20$, then divide both sides by 5 to get $y = \$4.00$. If the sirloin costs $1.50 more than the chuck, it costs $\$1.50 + \$4.00$ or $\$5.50/\text{pound}$.

21. **EXPLANATION: The correct answer is 9, 10, 11, 12, 13, or 14.**

Any of these would be credited as correct. Each chain costs $20 and each seat costs $30. That means the total amount, in dollars, is $20x + 4y$. Because the Homeowners Association spends at least $300 but no more than $400 on swing repairs, one can write the compound inequality $20x + 4y \geq \$300$ and $20x + 4y \leq \$400$. Because y is given as $30, we can substitute this value in the inequalities. $20x + 4(30) \geq 300$ and $20x + 120 \leq 400$. Subtract 120 from each side of both inequalities and divide by 20. This yields $x \geq 9$ and $x \leq 14$. Thus, x must be a whole number that is both greater than or equal to 9 and less than or equal to 14, but you can use any of the integers that fit that criteria as your answer. First you need to set up the inequality:

$$300 \leq 20x + 120 \leq 400$$
$$-\quad \underline{-120 \qquad\quad -120 - 120}$$
$$180 \quad \leq 20x \ \leq 280$$
$$9 \quad \leq \ \ x \ \leq 14$$

Therefore, x will be a whole number greater than or equal to 9 and less than or equal to 14. So grid in one of these: 9, 10, 11, 12, 13, 14. This is appropriate practice but not a system of equations.

22. **EXPLANATION: Choice C is correct.**

Build a system of equations from the facts provided in the problem. Use x = correct answers and y = incorrect answers. Since the total number of questions answered was 50, $x + y = 50$ states this correctly. Next, write a second equation to state the total score achieved. $70 = 2x - y$. Solve the system by the elimination method and divide by 3. The steps are as shown:

$$x + y = 50$$
$$\underline{2x - y = 70}$$
$$3x = 120$$
$$x = 40.$$

23. **EXPLANATION: Choice B is correct.**

To determine the cost of the car rental when it was equal to the cost of the SUV rental, determine the value of x (the number of days after Memorial Day) when the two prices were equal. This will happen when $S = C$—that is, when the two equations are equal. Set the equations equal, then simplify. First, subtract 75.50 from both sides: $10 + .15x = .25x$. Next, subtract $.15x$ from both sides: $10 = .10x$. Divide by .10 to solve for x. $x = 100$. Then to determine C, the cost of the car rental, substitute 100 for x in either equation, which gives both vehicle rentals at $100.50.

CHAPTER 7

LINES

On the SAT, you will need to write the **equation of a line** and find the slope and y-intercept from given points. You will also need to find the equation a line from a graph or perhaps because you know the line is either parallel or perpendicular to another line. There will be other questions your overall knowledge of lines. Begin with the mathematics review and then complete and correct the practice problems. There are 2 Solved SAT Problems and 21 Practice SAT Questions with answer explanations.

LINES

This chapter is about lines on a coordinate plane. Points on the coordinate plane (Cartesian plane) are named by ordered pairs (x, y).

There is an x-axis and a y-axis that cross at the Origin $(0, 0)$. The x-axis is horizontal, with positive values to the right of the Origin and negative values to the left. The y-axis is vertical with positive values above the Origin, and negative values below.

You will see graphs of the coordinate plane in the Examples below.

Example 1.

Find the slope-intercept equation of a line going through the points $(-2, 10)$ and $(3, -5)$.

The slope-intercept equation of a line is in the form $y = mx + b$, where m represents the slope and b represents the y-intercept.

First find the slope $m = \dfrac{Change\ in\ y}{Change\ in\ x} = \dfrac{y_2 - y_1}{x_2 - x_1} = \dfrac{-5 - 10}{3 - (-2)} = \dfrac{-15}{5} = -3$

Choose either point and the slope; substitute them into the equation to solve for b:

$y = mx + b \quad \Rightarrow \quad 10 = -3(-2) + b \quad \Rightarrow \quad 10 = 6 + b \quad \Rightarrow \quad 4 = b$

Or

$y = mx + b \quad \Rightarrow \quad -5 = -3(3) + b \quad \Rightarrow \quad -5 = -9 + b \quad \Rightarrow \quad 4 = b$

Therefore, the equation of the line is $y = -3x + 4$.

Example 2.

What is the equation of the line shown in the graph below?

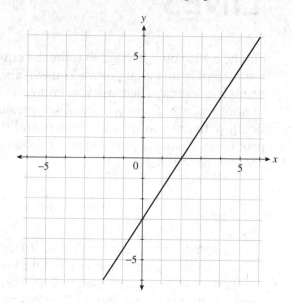

The graph has a y-intercept of -3; following the slope on the graph notice that the change in the y-value is 3 and the change in the x-value is 2. So the slope is $\dfrac{3}{2}$. Therefore, the equation of the line in the graph is $y = \dfrac{3}{2}x - 3$.

Example 3.

What is the slope of a line parallel to the line with equation $y = \dfrac{4}{7}x + 8$?

The slopes of parallel lines are the same; therefore the slope would be $m = \dfrac{4}{7}$.

Example 4.

What is the slope of a line perpendicular to the line with equation $y = \dfrac{5}{8}x - 6$?

The slopes of perpendicular lines are opposite reciprocals; therefore, the slope would be $m = -\dfrac{8}{5}$.

Example 5.

Find the equation of a line perpendicular to the line with equation $y = -\dfrac{2}{3}x + 9$ that passes though the point $(8, 5)$.

The slopes of perpendicular lines are opposite reciprocals; therefore, the slope would be $m = \dfrac{3}{2}$.

Substitute the point $(8, 5)$ and the slope $m = \dfrac{3}{2}$ into the equation to solve for *b*:

$$y = mx + b \quad \Rightarrow \quad 5 = \frac{3}{2}(8) + b \quad \Rightarrow \quad 5 = 12 + b \quad \Rightarrow \quad -7 = b$$

Therefore, the equation of the line is $y = \dfrac{3}{2}x - 7$.

Example 6.

Find the equation of a line parallel to the line with equation $6y - 8x = 18$ that passes though the point $(9, 4)$.

First rewrite the equation $6y - 8x = 18$ into slope-intercept form.

Add $8x$ to both sides of equation: $6y = 8x + 18$.

Divide both sides of equation by 6: $\dfrac{6y}{6} = \dfrac{8x}{6} + \dfrac{18}{6}$.

Simplify: $y = \dfrac{4}{3}x + 3$.

The slopes of parallel lines are the same; therefore the slope would be $m = \dfrac{4}{3}$.

Substitute the point $(9, 4)$ and the slope $m = \dfrac{4}{3}$ into the equation to solve for b:

$$y = mx + b \quad \Rightarrow \quad 4 = \frac{4}{3}(9) + b \quad \Rightarrow \quad 4 = 12 + b \quad \Rightarrow \quad -8 = b$$

Therefore, the equation of the line is $y = \dfrac{4}{3}x - 8$.

Practice Questions

1. Find the slope-intercept equation of a line going through the points $(6, -3)$ and $(8, 2)$.

2. What is the equation of the line shown in the graph below?

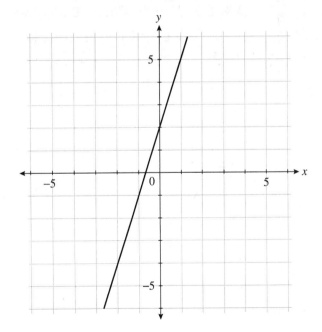

3. Find the equation of a line parallel to the line with equation $y = -\dfrac{7}{4}x - 11$ that passes though the point $(12, -9)$.

4. Find the equation of a line perpendicular to the line with equation $3y + 9x = 12$ that passes though the point $(6, -4)$.

Practice Answers

1. $y = \dfrac{5}{2}x - 18$.

2. $y = 3x + 2$.

3. $y = -\dfrac{7}{4}x + 12$.

4. $y = \dfrac{1}{3}x - 6$.

SOLVED SAT PROBLEMS

1. Some of the points on the graph of a linear equation are shown in the table below. Which of the following is the equation?

x	2	4	6
y	2	8	14

A. $y = 6x - 6$
B. $y = 3x - 4$
C. $y = 6x + 2$
D. $y = 3x + 2$

EXPLANATION: Choice B is correct.

Notice that the x-values increase by 2 and the y-values increase by 6, so the slope is

$$m = \frac{\text{Change in } y}{\text{Change in } x} = \frac{6}{2} = 3.$$

Choose one of the points and the slope; substitute them into the equation to solve for b. The point $(2, 2)$ is the easiest point to use and the slope is $m = 3$:

$$y = mx + b \quad \Rightarrow \quad 2 = 3(2) + b \quad \Rightarrow \quad 2 = 6 + b \quad \Rightarrow \quad -4 = b$$

Therefore, the equation of the line is $y = 3x - 4$.

2. Which of the following is the graph of the equation $y = -\dfrac{1}{2}x - 4$ in the xy-plane?

A.

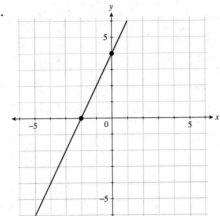

B.

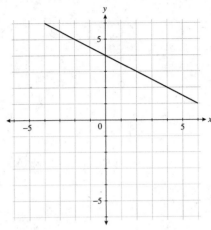

C.

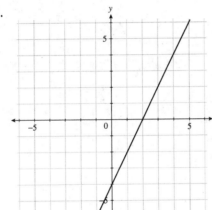

D.

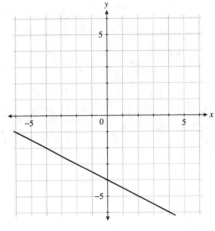

EXPLANATION: Choice D is correct.

Look for the graph with a y-intercept of $b = -4$ and a slope of $m = -\dfrac{1}{2}$. This occurs in choice D.

LINES
PRACTICE SAT QUESTIONS

ANSWER SHEET

Choose the correct answer.
If no choices are given, grid the answers in the section at the bottom of the page.

1. Ⓐ Ⓑ Ⓒ Ⓓ
2. Ⓐ Ⓑ Ⓒ Ⓓ
3. Ⓐ Ⓑ Ⓒ Ⓓ
4. Ⓐ Ⓑ Ⓒ Ⓓ
5. Ⓐ Ⓑ Ⓒ Ⓓ
6. Ⓐ Ⓑ Ⓒ Ⓓ
7. Ⓐ Ⓑ Ⓒ Ⓓ
8. Ⓐ Ⓑ Ⓒ Ⓓ
9. Ⓐ Ⓑ Ⓒ Ⓓ
10. GRID

11. Ⓐ Ⓑ Ⓒ Ⓓ
12. GRID
13. Ⓐ Ⓑ Ⓒ Ⓓ
14. Ⓐ Ⓑ Ⓒ Ⓓ
15. Ⓐ Ⓑ Ⓒ Ⓓ
16. Ⓐ Ⓑ Ⓒ Ⓓ
17. Ⓐ Ⓑ Ⓒ Ⓓ
18. GRID
19. Ⓐ Ⓑ Ⓒ Ⓓ
20. Ⓐ Ⓑ Ⓒ Ⓓ

21. Ⓐ Ⓑ Ⓒ Ⓓ

Use the answer spaces in the grids below if the question requires a grid-in response.

Student-Produced Responses

ONLY ANSWERS ENTERED IN THE CIRCLES IN EACH GRID WILL BE SCORED. YOU WILL NOT RECEIVE CREDIT FOR ANYTHING WRITTEN IN THE BOXES ABOVE THE CIRCLES.

10.

12.

18.

PRACTICE SAT QUESTIONS

1. A, B, and C are points on a line, and B is the midpoint of $\overline{AC}$. If $AB = 15$, then $AC =$
   A. 60
   B. 30
   C. 15
   D. 10

2. The equation of a line is $y = \dfrac{2}{3}x - 6$. What is the equation of a line perpendicular to this line and with the same y-intercept?

   A. $y = \dfrac{2}{3}x + 6$

   B. $y = \dfrac{2}{3}x - \dfrac{1}{6}$

   C. $y = \dfrac{3}{2}x - 6$

   D. $y = -\dfrac{3}{2}x - 6$

3. $-3x - 4y = 8$

   Which of the following equations is parallel to the equation above?
   A. $-3x - 4y = 12$
   B. $2x - 8y = 8$
   C. $4x + y = 3$
   D. $x + y = \dfrac{3}{4}$

4. A is the midpoint of $\overline{PQ}$ and B is the midpoint of $\overline{XY}$. $P = (2, 4)$ $Q = (6, 10)$ $X = (-8, 2)$, and $Y = (4, -6)$. What is the slope of $\overleftrightarrow{AB}$?

   A. $-\dfrac{2}{3}$

   B. $\dfrac{2}{3}$

   C. $\dfrac{3}{2}$

   D. $\dfrac{4}{7}$

5.

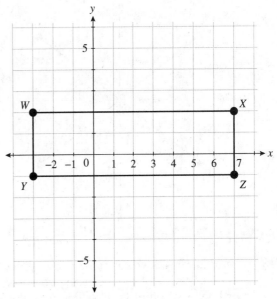

In the diagram above, $WXYZ$ is a rectangle. What is the equation of a line passing through W and Z?

   A. $y = -\dfrac{3}{10}x + 1.1$

   B. $y = \dfrac{3}{10}x + 1.1$

   C. $y = -\dfrac{3}{10}(x - 11)$

   D. $y = \dfrac{3}{10}x + 1.1$

6. Which of the following equations represents a line that is perpendicular to the line with equation $y = -\dfrac{1}{2}x + 23$?

   A. $-16x + 8y = 642$
   B. $10x - y = -11$
   C. $x + 3y = 12$
   D. $-\dfrac{1}{2}x + 2y = 14$

7. Points A, B, and C are colinear.
   If $\overline{AC} \cong \overline{BC}$ and $A = (-2, 3)$, $B = (2, -4)$, then $C =$

   A. $\left(0, -\dfrac{1}{2}\right)$

   B. $(0, -1)$

   C. $\left(-\dfrac{1}{2}, 0\right)$

   D. $\left(\dfrac{1}{2}, 0\right)$

8.

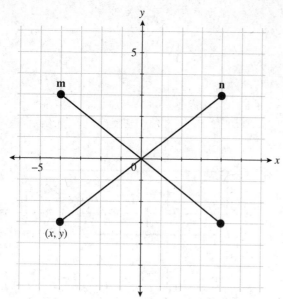

In the diagram above, line m is not perpendicular to line n. Which of the following could be the coordinates of point (x, y) if line m and line n were perpendicular?

A. $(-2, -5)$
B. $(-4, -3)$
C. $(-5, -2)$
D. $(2, 5)$

9. If $f(x)$ is a linear function such that $f(2) = 5$ and $f(4) = 13$, $f(x) =$

A. $f(x) = 3x - 4$
B. $f(x) = 4x - 3$
C. $f(x) = 4x + 3$
D. $f(x) = \dfrac{1}{4}x + \dfrac{9}{2}$

10. The slope of the line below is $\dfrac{3}{2}$. What is the y-intercept?

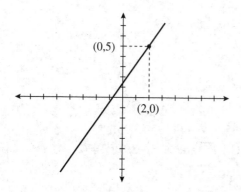

11. Points $R = (6, 4)$, $Q = (4, 3)$, and $P = (-2, b)$ are colinear. What is the value of b?

A. 4
B. 3
C. 2
D. 0

12.

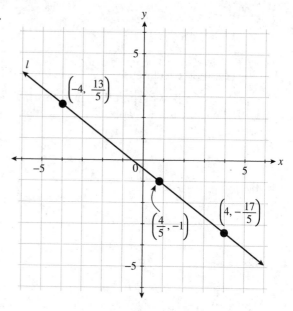

What is the negative of the slope of line l?

13. Which of the following graphs could be the graph of $y = 3x - 2$?

A.

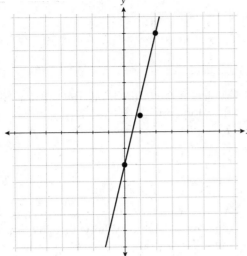

D.

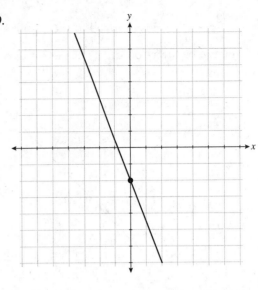

B.

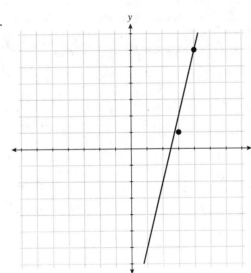

14. A line is plotted on the xy-plane with a point $(2, 3)$ in which the slope is equal to 2. What is the y-intercept?

A. −1
B. 0
C. 1
D. 2

15. A line on the coordinate plane has points at (a, b) and (c, d). If $c > a$ and $d > b$, then the slope of the line is:

A. not defined
B. zero
C. negative
D. positive

16. In the coordinate plane, which of the following points is on a line with a slope of $\frac{1}{8}$ that also passes through $(0, 0)$?

A. $(1, 8)$
B. $(2, 4)$
C. $(8, 4)$
D. $(16, 2)$

C.

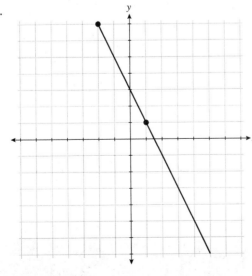

17. $g(x) = \frac{1}{4}x - b$

The value for $g(x)$ when $x = 4$ is 12. What is the value of $g(x)$ when $x = -12$?

A. −7
B. −5
C. 8
D. 10

18. If $\frac{2}{3}x + \frac{3}{4}y = 6$ is graphed in the xy-plane, what would be the x-coordinate of the x-intercept?

19.

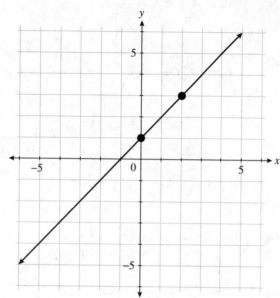

Which of the following is NOT an equation for the line on the graph above?

A. $x = y - 1$
B. $y = 1$
C. $y = x + 1$
D. $x - y = -1$

20. In the table below are some x and y values for a linear function.

x	y
1	6
2	8
3	10

Which of the following is the rule for this linear function above?

A. $x + 5$
B. $3x + 3$
C. $2x + 4$
D. $4x - 2$

21. A line in the coordinate plane passes through Quadrants III and IV only. Which of the following is the best description of the slope of the line?

A. The slope is positive.
B. The slope is negative.
C. The slope is 0.
D. The slope can be undefined.

EXPLAINED ANSWERS

1. **EXPLANATION: Choice B is correct.**

   B is the midpoint of $\overline{AC}$. $\overline{AC}$ is twice $\overline{AB}$. $AB = 15$.
   $AC = 2(15) = 30$.

2. **EXPLANATION: Choice D is correct.**

   Perpendicular lines have slopes that are opposite reciprocals. Therefore, the slope of the perpendicular line is $-\dfrac{3}{2}$

   The y-intercept is also -6 so, $y = -\dfrac{3}{2}x - 6$.

3. **EXPLANATION: Choice A is correct.**

   Find the slope of the equation by putting the original equation in $y = mx + b$ form where m is the slope: $-3x - 4y = 8$, and then move the x term so you have $-4y - 3x = 8$. Then, subtract the x term so you have $-4y = 3x + 8$. Divide both sides by 4, and the slope becomes $\dfrac{-3}{4}$. Equations of parallel lines always will have the same slope (m). Putting the original equation in choice A in standard form gives us $-4y = 3x + 12$. Dividing both sides by -4 gives $y = \dfrac{-3}{4}x - 3$, so that's the line where m is equal to the slope of the original equation $\left(\dfrac{-3}{4}\right)$.

4. **EXPLANATION: Choice C is correct.**

   Find the coordinates of A and B

   $$A = \left(\frac{2+6}{2}, \frac{4+10}{2}\right) = (4,7) \text{ and } B = \left(\frac{-8+4}{2}, \frac{2+(-6)}{2}\right) = (-2,-2)$$

   Use the coordinates to find the slope of $\overline{AB}$

   $$m_{\overline{AB}} = \frac{-2-7}{-2-4} = \frac{-9}{-6} = \frac{3}{2}.$$

5. **EXPLANATION: Choice A is correct.**

   In the xy-plane, the slope m of the line that passes through two points is the difference in the y-coordinates over the difference in the x-coordinates. Thus, the slope of the line through the points $w(-3, 2)$ and $z(7, -1)$ is $\dfrac{3}{-10}$. Thus, you can eliminate choices B and D. Plug a point into the remaining choices to see which one works. The equation of the line that passes through W and Z can be written $y = \dfrac{3}{-10}x + b$, where b is the y-intercept of the line. Since the line passes through W, the equation $2 = \dfrac{3}{-10}(-3) + b$ holds.

   This gives $2 = \dfrac{-9}{-10} + b$. If you subtract $\dfrac{-9}{-10}$ from the right side, you'll get $b = 2 - \dfrac{9}{10}$, which is 1.1. Therefore, the equation of the line that passes through points W and Z is $y = -\dfrac{3}{10}x + 1.1$.

6. **EXPLANATION: Choice A is correct.**

   Because the perpendicular slope of a line is the negative reciprocal, the slope we are looking for in the correct answer needs to be positive 2. Put each equation into $y = mx + b$ form. In this case, the equation in answer choice A becomes $8y = 16x + 642$. Divide everything by 8 to get $y = 2x + 80.25$: that's the correct slope, so you've found the answer.

7. **EXPLANATION: Choice A is correct.**

$\overline{AC} \cong \overline{BC}$, and they are colinear. That means C is the midpoint of $\overline{AB}$

$$C = \left(\frac{-2+2}{2}, \frac{3+(-4)}{2}\right) = \left(0, -\frac{1}{2}\right).$$

8. **EXPLANATION: Choice A is correct.**

Because (x, y) is in the third quadrant, it must have a negative x value and a negative y value so eliminate answer choice D. Next, use the formula to find the slope of line m: $\frac{y2-y1}{x2-x1}$, or in this case $\frac{-3-(+3)}{4-(-4)} = -\frac{6}{8}$, which reduces to $-\frac{3}{4}$. The slopes of perpendicular lines are negative reciprocals so the slope of line n must be $\frac{4}{3}$. Use the slope formula choices A, B, and C to see which choice gives you the correct slope value. Choice A works because $\frac{3-(-5)}{4-(-2)}$ equals $\frac{8}{6}$, which reduces to $\frac{4}{3}$.

9. **EXPLANATION: Choice B is correct.**

– Find the slope.
– Find the y-intercept.
– Substitute the slope and the y-intercept in the equation for the line and solve.

$$m = \frac{13-5}{4-2} = \frac{8}{2} = 4$$

$$f(x) = mx + b \Rightarrow f(x) = 4x + b$$

$$f(2) = 4(2) + b$$

$$5 = 4(2) + b \Rightarrow 5 = 4(2) + b$$

$$5 = 8 + b \Rightarrow -3 = b$$

$$f(x) = 4x - 3.$$

10. **EXPLANATION: The correct answer is 2.**

The slope of the line is $\frac{3}{2}$, and the point $(2, 5)$ is on the graph.

Substitute $m = \frac{3}{2}$, $x = 2$, $y = 5$ in the equation for a line. Solve for b.

$$y = mx + b \Rightarrow 5 = \frac{3}{2}(2) + b \Rightarrow 5 = 3 + b \Rightarrow 2 = b.$$

11. **EXPLANATION: Choice D is correct.**

First find the slope of $\overleftrightarrow{RQ}$.

$$m_{\overline{RQ}} = \frac{3-4}{4-6} = \frac{-1}{-2} = \frac{1}{2}$$

R, Q, and P are colinear, so the slope of $\overleftrightarrow{RQ}$ equals the slope of $\overleftrightarrow{RP}$. That means,

$$\frac{4-b}{6-(-2)} = \frac{4-b}{8} = \frac{1}{2} \Rightarrow 8 - 2b = 8$$

$$-2b = 0 \Rightarrow b = 0.$$

12. **EXPLANATION: The correct answer is $\frac{3}{4}$.**

    Use any two points on the graph and the slope formula to calculate the slope. First find the difference of the y-coordinates and divide it by the difference of the x-coordinates. The slope $= -\frac{3}{4}$, but the question asks for the negative of the slope so grid in $\frac{3}{4}$.

13. **EXPLANATION: Choice A is correct.**

    This equation is in slope-intercept from ($y = mx + b$), and gives the slope (m) as 3 and the y-intercept as –2. A positive slope goes up from left to right, which eliminates choices C and D. Only graph A show a y-intercept of –2, so that must be the correct answer.

14. **EXPLANATION: Choice A is correct.**

    Plug the numbers into $y = mx + b$: $3 = 2(2) + b$. Then multiply the right side of the equation to get $3 = 4 + b$. Subtract 4 from each side of the equation, so $-1 = b$.

15. **EXPLANATION: Choice D is correct.**

    Plug in numbers if it helps you to visualize this problem better. The slope of the line is $\frac{d-b}{c-a}$. Both the numerator and the denominator, if you follow the rules given in the problem, must be positive: therefore, the slope of the line must also be positive.

16. **EXPLANATION: Choice D is correct.**

    Don't forget your slope formula! Change in y over change in x means that a y-coordinate of 1 and an x-coordinate of 8 would be a point on the line. Therefore, you are looking for a ratio of $\frac{1}{8}$ with 1 as your y-coordinate and 8 as your x-coordinate. Since $\frac{2}{16} = \frac{1}{8}$, your y coordinate would be 2 and your x-coordinate would be 16, which is answer choice D.

    You can also solve this problem by plugging in choices into the formula $y = mx + b$. Since this line passes through the origin, you can simplify the problem to $y = mx$. The slope is stated as $\frac{1}{8}$ and therefore the equation simplifies as follows: $y = \frac{1}{8}x$ or $x = 8y$. When answer choice D is plugged into the equation $(16, 2)$, you get $16 = 8(2)$, which is correct, so D is the correct answer. If you mixed up your x- and y-coordinate, you may have chosen choice A, which has a y-coordinate of 8 and an x-coordinate of 1 and is therefore the opposite of what you are looking for. The other choices are also incorrect because they do not satisfy the y-coordinate of 2 (or 1) that is needed for this equation.

17. **EXPLANATION: Choice C is correct.**

    Substitute 4 for x and 12 for $g(x)$ as the question indicates: $g(x) = \frac{1}{4}x - b$

    Solve: $\quad 12 = \frac{1}{4}(4) - b$

    $\quad 12 = 1 - b$

    Subtract 1 from both sides: $\quad 11 = -b$

    Divide both sides by –1: $\quad -11 = b$

    Plug that back into the original equation: $\quad g(x) = \frac{1}{4}(-12) - (-11)$

    $\quad G(x) = (-3) - (-11) = 8.$

18. **EXPLANATION: The correct answer is 9.**

The x-intercept will have a y-coordinate of (0). Substitute (0) for y. This gives $\frac{2}{3}x + \frac{3}{4}(0) = 6$; this simplifies to $\frac{2}{3}x = 6$. Multiply both sides of the equation by 3 to remove the denominator. $2x = 18$; $x = 9$. The x-intercept is (9, 0) and the coordinate the question is asking for is 9.

19. **EXPLANATION: Choice B is correct.**

Using the coordinates from the given graph, or any other two points, arrive at the slope of the line using the slope formula; (2, 3) and (0, 1) are good points to use, because one of them is also the y-intercept. The slope $= \frac{3-1}{2-0} = 1$, when simplified. You have the y-intercept as well. Put these facts into the slope/intercept form of the equation of the given line: $y = x + 1$. The only equation not equal to this one is B. The others are valid versions of this equation.

20. **EXPLANATION: Choice C is correct.**

The y values are all even, which means the answer must involve multiplying by an even number. Rule out choices A and B. Then, try out C and D by inserting the points into the expressions. Choice C works: $2(1) + 4 = 6$ for every set of values in the table, whereas D does not. Therefore, C is the correct answer.

21. **EXPLANATION: Choice C is correct.**

A line that crosses through Quadrants III and IV only has to be parallel to the x-axis, in which case the slope would be 0. Because it could potentially be a straight line, we can't be sure if it has a negative or positive slope, so A and B can't be the correct answers.

CHAPTER 8

LINEAR EQUATIONS AND WORD PROBLEMS

On the SAT, you will find the **equation of a line** in the context of a problem and solve or evaluate values using the equation. You will need also match graphs to applicable scenarios and interpret the meaning of the **slope** and the **y-intercept of the line** in the context of a problem. (The slope of a line represents the change in y over change in x $\frac{y2 - y1}{x2 - x1}$.) The y-intercept of a line is the point at which the line crosses the y-axis where x is equal to zero. Begin with the mathematics review and then complete and correct the practice problems. There are 2 solved SAT Problems and 25 Practice Questions with answer explanations.

LINEAR EQUATIONS AND WORD PROBLEMS

Here are some of the types of problems you will find on the SAT.

Point-Slope Form

Use the Point-Slope form primarily to write the equation of a line when you know two points on the line. (1) Use the points to find the slope. (2) Substitute coordinates from one of the points and the slope in the point-slope formula, which is $y - y_1 = m(x - x_1)$. (y_1 and x_1 are the coordinates of one of the known points.)

Example 1.

The stopping distance of a car traveling 30 miles per hour is 115 feet and the stopping distance of a car traveling 40 miles per hour is 165 feet. Assuming the stopping distance of a car increases at a constant rate based on its speed, write the equation for stopping distance, D, in terms of speed, S.

We know two points where x is speed and y is distance: (30, 115) and (40, 165).

Use the slope formula $\frac{y2 - y1}{x2 - x1}$ to determine the slope of the line: $\frac{(165 - 115)}{(40 - 30)} = \frac{50}{10} = 5$.

Now, use point-slope form, which is $y - y_1 = m(x - x_1)$. You can use either of the points.

$y - 165 = 5(x - 40)$. Now simplify:

$y - 165 = 5x - 200$

$y = 5x - 35$. Our x was the speed and y was the distance, so our equation is

$D = 5S - 35$.

Slope

Example 2.

The relationship between the average weight of an orange on an orange tree and the number of oranges on the tree can be modeled by the equation $y = -0.0256x + 0.6$, where x represents the number or oranges on the tree, and y represents the average weight per hundred oranges in pounds. Give a proper interpretation of the number -0.0256 in the equation.

> The slope in word problems usually represents the rate of change. In this case, the rate of change is negative and as the number of oranges goes up, the average weight per hundred oranges goes down, by a factor of 0.0256 pounds.

y-intercept

Example 3.

The average number of minutes an online shopper age 17 and over is on a specific website each month can be modeled by the equation, $y = -11.5x + 554$, where x represents the age of the person in years who is at least 17 and y represents the average number of minutes on the website each month. Give a proper interpretation of the number 554 in the equation.

> In this equation, 554 is the starting value of minutes a 17-year-old shopper is on the website, and it decreases from there depending on how much older than 17 the shopper is. The y-intercept of an equation usually represents a starting value.

Solving the Equation of a Line

Example 4.

Derek needs to rent a truck. The truck rental company charges $25 for the first 2 hours and then $15 for each additional hour, and Derek has $80 to spend on the truck rental. If the company only rents the truck for a whole number of hours, what is the maximum number of hours Derek can rent the truck?

> You can use an equation similar to the ones we've been working with in the previous examples. $25, the charge for the first two hours, is your y-intercept, while $15 is the rate of change (in this case, dollars per hour), so the equation would be $y = 15x + 25$ to represent the $25-dollar flat fee and then the $15 per hour.

> To solve, set the equation equal to 80.

> $80 = 15x + 25$

> $55 = 15x$

> $x = 3.67$, which rounds down to 3, since the truck cannot be rented for 0.67 of an hour, and rounding up to 4 would increase the cost over $80. Don't forget that x is the number of additional hours, and doesn't include the first two hours. The maximum number of hours Derek can rent the truck is thus $3 + 2 = 5$ hours.

Interpreting the Graph of a Line

Example 5.

The y-axis on the graph below shows the average monthly price, in dollars per barrel, for crude oil in 2006. On the x-axis, 1 represents January and 12 represents December). Between what two months did the average price of crude oil increase or decrease the slowest?

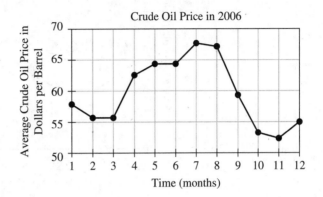

It can be hard to know precise numbers when looking at a graph. However, on this graph, between months 2 and 3 (FebruaryMarch) and months 5 and 6 (May/June) show the slowest increase or decrease. The slope of the two points would be the rate of growth.

Practice Questions

1. A 170 calorie hot dog that has contains 15 grams of fat and a 120 calorie hot dog contains 10 grams of fat. Assuming grams of fat and calories in a hot dog are linearly related, write an equation for the number of calories, C, in a hot dog with F grams of fat.

2. The mean error in nautical miles for landfall predication of a hurricane from 2000 to 2016 can be modeled by the equation $y = -2.03x + 67$, where x represents number of years since 2000 and y represents mean error in nautical miles for landfall. Give a proper interpretation of the number -2.03 in the equation.

3. The average speed of the Tour de France winner from 1960 to 2017 can be modeled by the equation $y = 0.108x + 34.9$, where x represents number of years since 1960 and y represents average speed of the Tour de France winner. Give a proper interpretation of the number 34.9 in the equation.

4. A car that averages 25 miles per gallon on the highway emits on average 446 grams of CO_2 per mile. For cars that average between 25 miles per gallon and 50 miles per gallon, the emission rate decreases on average by 13 grams of CO_2 per mile per gallon. What is the predicted emission rate of a car that averages 35 miles per gallon?

5. The graph below shows the yearly rainfall amount, in inches, for Los Angeles, California during the years 2012–2018. Between what two years did the yearly rainfall amounts increase at the slowest rate?

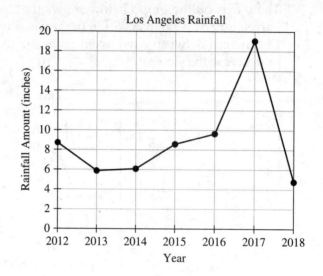

Practice Answers

1. $C = 10F + 20$.
2. The −2.03 represents the rate of change: the error is decreasing by 2.03 miles per year since 2000.
3. 34.9 is the starting value of the winner's average speed when the phenomenon being modeled by the equation starts: in this case, in 1960.
4. 316 g of CO_2/mile.
5. 2013–2014.

SOLVED SAT PROBLEMS

1. $C = 25h + 110$

    The equation above gives the amount, represented by C, in dollars, that an electrician charges for a job that takes h hours. Mr. Collins and Mr. Kennedy each hired this electrician. The electrician worked two hours longer on Mr. Collins's job than she did on Mr. Kennedy's job, which was less complex and required fewer new parts. How much more money did the electrician charge Mr. Collins than she charged Mr. Roland? Assume the additional parts do not need to be included in the calculation.

    A. $25
    B. $50
    C. $135
    D. $320

    EXPLANATION: Choice B is correct.

    One way to solve this question is by plugging in a number. Start with Mr. Kennedy's job, and say it took one hour: the cost of that job would be modeled by

    $25(1) + 110$ for a total cost, C, of $135.

    Now, for Mr. Collins's job, our number of hours is two more, according to the problem, so solve the same equation but plug in 3 instead. The cost of this job is modeled by

    $25(3) + 110$ for a total cost, C, of $185.

    The question is asking how much more Mr. Collins was charged, so subtract the two numbers: $185 - 135 = 50$.

2. From 2000 to 2014 the percentage of female smokers decreased from 24.9% to 14.8%. Assuming that the percentage of female smokers decreased at a constant rate, which of the following linear equations best models the percentage of female smokers x years after year 2000?

    A. $y = -0.72x + 24.9$
    B. $y = 0.72x + 24.9$
    C. $y = -0.72x - 24.9$
    D. $y = 0.72x - 24.9$

    Start by finding the slope of the line from two points. The points are $(0, 24.9)$, which represent the year 2000, and $(14, 14.8)$, which represent the year 2014. The slope is $\dfrac{24.9 - 14.8}{0 - 14} = \dfrac{10.1}{-14} = -0.72$.

    Rule out choices B and D because the slopes are positive. The y-intercept is 24.9, corresponding to choice A.

LINEAR EQUATIONS AND WORD PROBLEMS
PRACTICE SAT QUESTIONS

ANSWER SHEET

Choose the correct answer.
If no choices are given, grid the answers in the section at the bottom of the page.

1. (A)(B)(C)(D)	11. (A)(B)(C)(D)	21. (A)(B)(C)(D)
2. (A)(B)(C)(D)	12. (A)(B)(C)(D)	22. GRID
3. (A)(B)(C)(D)	13. (A)(B)(C)(D)	23. GRID
4. (A)(B)(C)(D)	14. (A)(B)(C)(D)	24. GRID
5. (A)(B)(C)(D)	15. (A)(B)(C)(D)	25. GRID
6. (A)(B)(C)(D)	16. (A)(B)(C)(D)	
7. (A)(B)(C)(D)	17. (A)(B)(C)(D)	
8. (A)(B)(C)(D)	18. (A)(B)(C)(D)	
9. (A)(B)(C)(D)	19. (A)(B)(C)(D)	
10. (A)(B)(C)(D)	20. (A)(B)(C)(D)	

Use the answer spaces in the grids below if the question requires a grid-in response.

Student-Produced Responses	ONLY ANSWERS ENTERED IN THE CIRCLES IN EACH GRID WILL BE SCORED. YOU WILL NOT RECEIVE CREDIT FOR ANYTHING WRITTEN IN THE BOXES ABOVE THE CIRCLES.

22.

23.

24.

25.

PRACTICE SAT QUESTIONS

Questions 1 and 2 are based on the information below:

A production line has an error rate based on the number of workers on the line. The formula for production with no errors is $p = 10{,}670 - 600w$, where w is the number of workers on the line and p is the production rate without errors.

1. Which of the following expresses this formula in terms of w?

    A. $w = \dfrac{p - 10{,}670}{600}$

    B. $w = p + (10{,}670 - 600)$

    C. $w = \dfrac{10{,}670 - p}{600}$

    D. $w = \dfrac{p + 600}{10{,}670}$

2. At which number of workers will the production line have an error-free production rate of 8,870?

    A. 1
    B. 2
    C. 3
    D. 4

3. A business delivers packages in a rental truck. The table below shows the cost of the packages regularly delivered on Monday, Wednesday, and Friday, along with the daily cost of the truck rental and the daily cost of insurance.

Day of the Week	Package Cost (P)	Daily Truck Rental (R)	Daily Insurance (I)
Mondays	$1,200	$80	$100
Wednesdays	$1,000	$100	$120
Fridays	$1,400	$90	$110

The total cost (T) of delivering packages each day is $T = P + (R + I)$.

What is the fewest number of days (D) that the total of package cost (P), truck rental (R), and insurance (I) on Wednesdays is greater than or equal to that cost of three Monday deliveries?

A. 2
B. 3
C. 4
D. 5

4. A catering company estimates the cost of catering an event using a formula they have come up with, $7{,}300 + 13bh$. b is the number of employees that will be working at any given event, and h is the time the job will take using those employees and no others, expressed in hours. In the formula used by the catering company, why is the expression bh multiplied by 13?

    A. The catering company charges $13 an hour for each employee.
    B. A maximum of 13 employees will work on the event.
    C. The price of an event increases $13 each hour the event goes on.
    D. Each event is 13 hours in length.

5. A baker is going to bake a certain number of cupcakes with the same shape and size as her main competitor. She calculates her costs using the following formula: $pX + 5p$. p is the number of cupcake wrappers needed, and X is a variable representing dollars per total unit cost for ingredients. Then she adds on a flat fee of $5 per cupcake to make a profit. If the baker decides to go with less expensive ingredients, which of the parts of the formula changes?

    A. p
    B. X
    C. 5
    D. $5p$

6. Earth's atmosphere is made up of chemical elements. The table below shows the weight of three elements in a cubic foot of air, rounded to the nearest hundredth.

	Weight of Elements in a Cubic Foot of Air	
	Grams	Ounces
Nitrogen (N)	27.5	0.96
Oxygen (O)	8.5	0.3
Argon (A)	0.15	0.01

If x ounces of an element are equal to y grams of an element, what is the relationship between x and y?

A. $y = 0.134x - 28.78$
B. $y = 0.134x + 28.78$
C. $y = 28.78x - 0.134$
D. $y = 28.78x + 0.134$

7. $b + w = 3.2$

This equation represents Liliya's language study plan before she goes to study abroad for the summer. It shows the hours, represented by b, she spends studying vocabulary each day and the amount of hours, w, she spends studying grammar each day. What does the number 3.2 indicate here?

A. The number of hours spent studying grammar each day
B. The number of hours spent studying vocabulary each day
C. The total number of hours spent studying language each day
D. The number of minutes spent studying grammar for each minute spent studying vocabulary

8. A bakery owner decides to increase the bakery's cookie offerings. Starting with 12 types of cookies, the bakery will add two new offerings each week for the next 10 weeks. The total number of offerings is represented by c, and the number of weeks is represented by w. If the equation $c = mw + b$ represents the relationship, which variable should be replaced with the number 12?

A. c
B. w
C. m
D. b

9. Lydia owns a cupcake bakery in which she is the primary baker as well as the owner. The number of cupcakes she has left to bake until the end of her shift is calculated at any given time in the shift using the formula $N = 94 - 12e$. In this formula, N is the number of cupcakes Lydia has left to make and e is the number of hours she has worked so far that shift. What does 94 represent?

A. Lydia will bake all the cupcakes she has orders for in 94 hours of work.
B. Lydia starts each shift with 94 cupcakes to bake.
C. Lydia bakes cupcakes at a rate of 94 cupcakes per hour for all the hours of her shift.
D. Lydia bakes cupcakes at a rate of 94 cupcakes in total per shift she completes.

10. A baseball card collector buys and sells cards with the image of Lou Gehrig, a famous baseball player. The graph below shows how the number of cards in the collection increases and decreases each month. During which period did the collection increase the most?

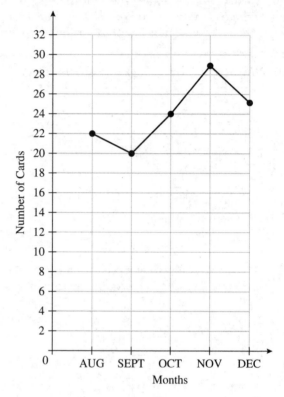

A. August to September
B. September to October
C. October to November
D. November to December

11. House construction began at a good pace. Then the construction supervisor noted a problem and parts of the construction had to be removed. After that, things leveled off before beginning again at a faster pace than during the first phase. Which of the following graphs best depicts the construction process?

A.

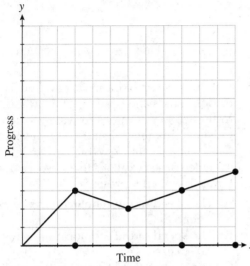

B.

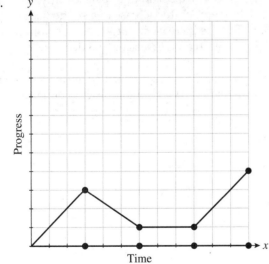

C.

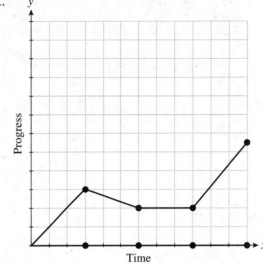

D.

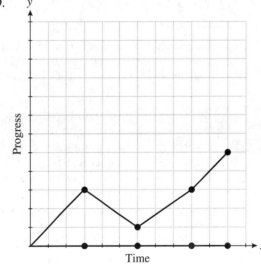

12. A small boat operates between the mainland and the beach. The boat departs on the hour from 9:00 a.m. to 1:00 p.m. Based on the graph below, which would be a correct statement about the number of passengers the boat carried?

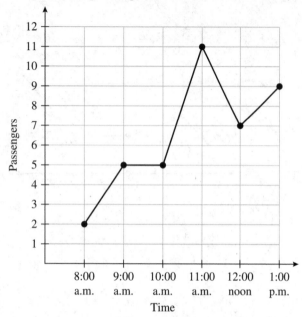

A. The number of passengers on the boat steadily increased throughout the day.
B. The boat carried the same number of passengers on two consecutive trips.
C. The most passengers were between noon and 1:00 p.m.
D. There were more passengers at 9 a.m. than at noon.

13. The machine shop used the formula below to determine how many centimeters, c, a screw was driven into a hole based on the number of screw turns, t.

$$c = \frac{10t}{100}$$

Based on the equation, how much further was the screw driven into the hole for each turn?

A. $\frac{1}{10}$ cm

B. $\frac{2}{10}$ cm

C. $\frac{1}{100}$ cm

D. $\frac{2}{100}$ cm

14. $p = 2g + 25.4$

A teacher uses the formula above to estimate the reading progress, p, of a student, in terms of the student's grade level, g. This formula is valid, according to the teacher, for students between grades of 2 and 4. What is the estimated progress number of a typical student each year?

A. 2
B. 5.1
C. 8.46
D. 12.7

15. Dolphins use echolocation to navigate. The speed of dolphins' echolocation, D, in water, relates to the temperature of the water, T, in degrees Celsius. This relationship can be expressed by $D(T) = 0.08T + 442.5$. Which of the following statements best explains the presence of 442.5 in terms of what it represents?

A. The speed of the echolocation waves at 0 degrees C
B. The speed of the echolocation waves at 0.08 degrees C
C. The increase in the speed of the echolocation waves that happens over the course of a temperate increase of a single degree
D. The increase in the speed of sound that goes with an increase of 0.08 degrees C

16. It costs 25 cents for each 10 minutes at a parking meter. Which of the following equations shows how much it costs (c) in dollars for h hours at the meter?

A. $c = 0.25(h)$
B. $c = 0.25(60h \div 10)$
C. $c = 0.25/10(h)$
D. $c = 0.25(h \div 10)$

17. Occasionally, it is helpful to convert Celsius temperature to Fahrenheit temperature. The formula is most often used to make this conversion

    $$F = C \times \frac{9}{5} + 32$$

    Which of the following is a correct interpretation of the formula?

    I. The equivalent of 0 degrees C is 32 degrees F.

    II. To convert from C to F, multiply by $\frac{9}{5}$ and add 32.

    III. To convert from F to C, divide by $\frac{5}{9}$ and subtract 32.

    A. I only
    B. III only
    C. I and II
    D. II and III

18. John, a garden designer, uses a formula $b = 12ac$ to estimate the number of seeds, b, that he will need to use in order to completely fill with plantings a flowerbed. The bed measures a feet wide and c feet long. Which of the following gives c in terms of b and a?

    A. $c = \dfrac{12}{ab}$

    B. $c = \dfrac{a}{12b}$

    C. $c = \dfrac{b}{12a}$

    D. $c = \dfrac{b}{12 + a}$

19. Angela wants to buy tickets for a symphony concert that is sold out. She finds some from a ticket reseller, who will charge her a service fee. The equation $y = 121x + 23$ represents the total amount, y, that Angela will have to pay to buy x tickets. What does 121 mean?

    A. The price of a single ticket
    B. The amount of the service fee being charged
    C. The total amount Angela will pay if she buys just one ticket.
    D. The total amount Angela will pay for any number of tickets she chooses to buy

20. A company that tests recipes buys equipment that costs a total of $20,500 when they move into their new kitchen space. The equipment depreciates consistently for 5 years, when it is considered to have no value to the company and is no longer listed among company assets. What is the salvage value of the equipment after 3 years from when the company moved into their new studio?

    A. $4,100
    B. $8,200
    C. $12,300
    D. $16,400

21. Lucy wants to rent a popcorn machine for a backyard movie night. The machine costs $50 per hour, and she also needs to buy popcorn and toppings that cost $27 in total. Lucy has budgeted $315 for the movie night and has no other expenses, since she already owns the projector, screen, and movie she plans to show. How many hours can Lucy rent the machine for without going over her budget?

    A. 3
    B. 4
    C. 5
    D. 6

22. A ship increases speed each hour at a constant rate as it moves away from shore. The ship's speed is recorded in knots. At the fifth hour, the ship's speed was 11.5 knots. At the twelfth hour the speed was 20.6 knots. How much did the ship's speed increase each hour?

23. It costs Derek $10 to enter the carnival. Then it costs an additional $7 for each attraction. Derek started with $100 and spent the maximum amount possible on attractions. How many attractions did Derek see?

24. A race car starts out around a track at a relatively low speed of 140 mph. The race car increases speed by 7 miles per hour each lap after that. What is the the speed of the race car on the fourth lap?

25. Bob belongs to a comprehensive health club that includes a gym and classes. It costs Bob $59.95 a month to belong to the gym. Classes are $12.50 each. In one month, Bob spent $134.95 on dues and classes. How many classes did Bob pay for?

CONQUERING SAT MATH

EXPLAINED ANSWERS

1. **EXPLANATION: Choice C is correct.**

"In terms of w" is your clue that the problem wants you to isolate w on one side of the equation. In the equation $p = 10{,}670 - 600w$, add $600w$ to both sides to produce a positive "w": $p + 600w = 10{,}670$. Subtract p from both sides to begin isolating w: $600w = -p + 10{,}670$. Then divide both sides by "600":

$$\frac{600w}{600} = \frac{-p + 10{,}670w}{600} = \frac{-p + 10{,}670}{600}$$

Answer C is the equivalent to this equation.

2. **EXPLANATION: Choice C is correct.**

Use the original formula and solve for w. Substitute 8,870 for p. $8{,}870 = 10{,}670 - 600w$. Subtract 10,670 from both sides: $-1{,}800 = -600w$, and then divide both sides by -600 to get $3 = w$.

3. **EXPLANATION: Choice C is correct.**

Use the information from the table and the formula. The question asks for the fewest number of days, among the choices given, that Wednesday's costs are ≥ 3 times Monday's costs. We don't know the number of Wednesday's deliveries, so use D and Wednesday's costs are $D(1{,}000 + 100 + 120)$. We know there are three Monday deliveries. So, Monday's costs are $3(1{,}200 + 80 + 100)$.

Write an inequality: $D(1{,}000 + 100 + 120) \geq 3(1{,}200 + 80 + 100)$. Simplify each side: $1{,}220 D \geq 3(1{,}380)$. $1{,}220 D \geq 4{,}140$. $D \geq 3.39$. 4 is the smallest number of days greater than 3.39. That's choice C.

4. **EXPLANATION: Choice A is correct.**

The 7,300 is the fixed cost, and the 13 is the amount that the cost will grow by per hour and employee (since it is multiplied by both those variables). Therefore, the expression bh is multiplied by 13 to consider both the number of hours and the number of employees. Only Choice A represents this.

5. **EXPLANATION: Choice B is correct.**

If the baker uses less expensive ingredients, her cost will change (decrease). This is represented by the X variable in the formula. Be sure you read this type of problem carefully so you don't miss anything!

6. **EXPLANATION: Choice C is correct.**

Use two points to calculate the slope first: (0.3, 8.5) and (0.96, 27.5). Using the slope formula, which is the difference in y over the difference in x, you get $\frac{19}{0.66}$, which simplifies to 28.78: this is the slope. Then, use the point-slope formula: $y - 8.5 = 28.78(x - 0.3)$. Simplify: $y - 8.5 = 28.78x - 8.634$. Then, add 8.5 to both sides to get choice C. Alternatively, test your answer choices by plugging in a point. Make sure to remember that x is ounces!

7. **EXPLANATION: Choice C is correct.**

In this equation, the 3.2 is the total that the two variables, representing the times for the different studying topics, add up to. Therefore, it must be the total time each day.

8. **EXPLANATION: Choice D is correct.**

12 is the original number of varieties of cookies, the starting value. That is represented by the y-intercept of the equation, which is in the b slot.

9. **EXPLANATION: Choice B is correct.**

94 is being decreased by 12 cupcakes for each hour, e, that Lydia works, so the 94 must be the number of cupcakes she started with.

10. **EXPLANATION: Choice C is correct.**

    You might be able to do an easy version of this question by looking, but don't count on it! The safest way to ensure you get the question right is to use the slope formula to determine the slope of the line between each set of two points. In this case, the points can be (3, 24) using the 3 to represent October as the third month on the graph and (4, 29), since November is the fourth month on the graph. That slope is steeper than any other on the graph, so that's the answer.

11. **EXPLANATION: Choice C is correct.**

    Use one piece of the problem at a time—we need a slow rise, a dip, and then a flat period, followed by a steeper rise in the line over time. Only choice C meets those requirements.

12. **EXPLANATION: Choice B is correct.**

    The lack of change in number of passengers over time indicates that the number of passengers stayed the same, even though the hour moved from one to the next. This matches choice B.

13. **EXPLANATION: Choice A is correct.**

    Plug 1 in for t to determine how many centimeters of change you have for one full turn.

14. **EXPLANATION: Choice A is correct.**

    2 is the slope in this equation, so it is the rate of change per year. Therefore, the rate of change can be described as a change of 2 reading levels for every school year.

    If you're unsure, plug in 1 for g, then p becomes $p = 2g + 25.4$. Substitute 1 for g, so $p = 2(1) + 25.4 = 2 + 25.4 = 27.4$. Therefore, the formula shows that the student's reading level increases by 2 levels each year.

15. **EXPLANATION: Choice A is correct.**

    The 442.5 is the starting value of the speed, before it begins to vary based on the temperature. You could also think of it as the y-intercept of the equation, or where $x = 0$, which is what choice A is saying.

16. **EXPLANATION: Choice B is correct.**

    The facts in the problem give cost (c) for minutes but states the time as hours (h). Time in hours (h) must be changed to minutes to calculate the cost (c). The equation should show the number of 10-minute time groups times 0.25. That is choice B.

17. **EXPLANATION: Choice C is correct.**

    II is an easy one to start with here—it's clearly true, since it is just a word-based version of the formula above with C isolated instead of F. That allows us to get rid of answer choices A and B, and then you only need to check either I or III, whichever makes more sense to you. I used the fact that the 32 is the y-intercept of the equation, or starting value, whereas III is a misrepresentation of the relationship given.

18. **EXPLANATION: Choice C is correct.**

    To isolate c, divide away all the other variables ($12a$) on both sides: $b = 12ac$ becomes $\dfrac{b}{12a} = C$. Doing so gets you choice C, $c = \dfrac{b}{12a}$.

19. **EXPLANATION: Choice A is correct.**

    Mathematically, the 121 is the rate of change for Angela's spending (slope). Therefore, the total cost (y) is the price of a ticket multiplied by the number of tickets (x) plus a service fee (which is represented by the number 23 in the problem). Therefore, she's going to spend $121 per ticket —that is choice A. You can solve this problem by thinking about the equation, which has a per-ticket cost of $121 and a one-time service fee of $23. The per ticket cost is Choice A.

20. **EXPLANATION: Choice B is correct.**

Think of the equation as a straight line depreciation problem: In order to solve this problem, set up an equation as follows: (Cost) − (Salvage value) = Amount depreciated, where the annual depreciation equals (Amount depreciated)/(Number of years). Then solve:

$20,500 − 0 = $20,500   Find the amount depreciated.

$20,500/5 = $4,100   Find the annual depreciation.

3 × $4,100 = $12,300   Find the depreciation after 3 years.

$20.500 − $12,300 = $8,200   Find the salvage value after 3 years.

You can also solve this problem by creating two points from this equation: (0, 20,500) for when the equipment is bought, and (5, 0) for when it has no value. Then, write an equation. Finding the slope using the slope formula should give you −4,100. Use point-slope now: $y − 0 = −4,100 (5 − x)$, then $y = −20,500 + 4,100x$. Plug in 3 for x, the number of years, then solve: $y = −20,500 + 4,100(3)$, then $y = −20,500 + 12,300$. $y = 8,200$. Don't forget that any problem can be solved any way you're comfortable: the important thing is getting it right!

21. **EXPLANATION: Choice C is correct.**

Some problems are more easily solved in a concrete way than by writing equations, so don't be afraid to do so. In this case, it's easy to subtract the money for the popcorn and toppings ($27) from the $315, which gives you $288 and then divide by $50 to get 5.76, or 5 hours since it must be rounded down to a whole number. If you'd prefer to do the problem algebraically, the equation would be $315 = 50h + 27$, where h is the number of hours.

22. **EXPLANATION: The correct answer is 1.3.**

This can be solved with an equation by using the points (5, 11.5) and (12, 20.6). The slope, which is all that is needed, is 9.1 divided by 7, $\left(\dfrac{9.1}{7}\right)$, or 1.3 knots.

23. **EXPLANATION: The correct answer is 12.**

For this one, it may be easiest to work backwards using numbers. Subtract the $10 entry fee, then divide the $90 remaining by $7 per attraction: you'll get 12.8, which means he could afford 12 attractions.

24. **EXPLANATION: The correct answer is 161.**

The easiest way to do this question is to simply add 7 miles an hour for each lap after the first lap. The race car begins at 140 mph for the first lap, so the second is at 147 mph, the third at 154 mph, and the fourth at 161 mph Some people will use this expression: $(3 × 7) + 140 = 21 + 140 = 161$.

Note that we couldn't just multiply 7 times 4 and add to 140 mph, since the first lap was itself at 140 mph. Don't be afraid to use a simple method to solve a complex problem!

25. **EXPLANATION: The correct answer is 6.**

Like the problems right before it, you can choose to solve this with an equation or with a more concrete method. If you choose to do it with an equation, the equation in question would be $134.95 = $12.50c + 59.95, where c is the number of classes.

CHAPTER 9
INEQUALITIES

On the SAT, you will need to find solutions to **inequalities** and **systems of inequalities**. An inequality compares two values to show their relationship. A value is either equal to (=), less than to (<), greater than (>), less than or equal to (≤), greater than or equal to (≥), equal another value. You will also need to be able to identify inequalities and systems of inequalities that describe a given situation. Begin with the mathematics review and then complete and correct the practice problems. There are 2 Solved SAT Problems and 14 Practice Questions with answer explanations.

INEQUALITIES

Example 1.

Is 5 a solution to the inequality $3x + 30 < 45$?

Substitute 5 for x in $3x + 30 < 45$.

$$3(5) + 30 < 45 \quad \Rightarrow \quad 15 + 30 < 45 \quad \Rightarrow \quad 45 < 45$$

It is not true that $45 < 45$, and 5 is not a solution to the inequality $3x + 30 < 45$.

Example 2.

Is 14 a solution to the inequality $2x + 18 \leq 46$?

Substitute 14 for x in $2x + 18 \leq 46$.

$$2(14) + 18 \leq 46 \quad \Rightarrow \quad 28 + 18 \leq 46 \quad \Rightarrow \quad 46 \leq 46$$

Notice that $46 = 46$ so $46 \leq 46$, and 14 is a solution to $2x + 18 \leq 46$.

Example 3.

Is the point $(4, 2)$ a solution to the inequality $3x - 4y \leq 8$?

Substitute $x = 4$ and $y = 2$ into $3x - 4y$ and solve: $3(4) - 4(2) = 12 - 8 = 4$
Since $4 < 8$, the point $(4, 2)$ is a solution to the inequality $3x - 4y \leq 8$.

Example 4.

Is the point $(-2, -1)$ a solution to the inequality $2x - 5y > 1$?

Substitute $x = -2$ and $y = -1$ into $2x - 5y$ and solve: $2(-2) - 5(-1) = -4 + 5 = 1$
Since the solution is 1, and we are looking for solutions more than 1; the point $(-2, -1)$ is not a solution to the inequality $2x - 5y > 1$.

Example 5.

Is the point $(-2, 2)$ a solution to the system of inequalities shown below?

$x + 2y < 4$

$2x - 3y \geq 6$

Try the first inequality.

Substitute $x = -2$ and $y = 2$ into $x + 2y$ and solve: $(-2) + 2(2) = -2 + 4 = 2$

Since $2 < 8$, the point $(-2, 2)$ is a solution to the inequality $x + 2y < 4$.

Try the second inequality.

Substitute $x = -2$ and $y = 2$ into $2x - 3y$ and solve: $2(-2) - 3(2) = -4 - 6 = -10$

Since $-10 < 6$, the point $(-2, 2)$ is not a solution to the inequality $2x - 3y \geq 6$.

Since the point $(-2, 2)$ does not work for both inequalities, it is not a solution to the system of inequalities.

Example 6.

Is the point $(3, -4)$ a solution to the system of inequalities shown below?

$2x - y > 5$

$3x + 2y \leq 7$.

Try the first inequality.

Substitute $x = 3$ and $y = -4$ into $2x - y$ and solve: $2(3) - (-4) = 6 + 4 = 10$

Since $10 > 5$, the point $(3, -4)$ is a solution to the inequality $2x - y > 5$.

Try the second inequality.

Substitute $x = 3$ and $y = -4$ into $3x + 2y$ and solve: $3(3) + 2(-4) = 9 - 8 = 1$

Since $1 < 7$, the point $(3, -4)$ is a solution to the inequality $3x + 2y \leq 7$.

Since the point $(3, -4)$ works for both inequalities, it is a solution to the system of inequalities.

SOLVING AND GRAPHING INEQUALITIES

When solving an inequality you must switch the sign of the inequality if you multiply or divide by a negative number.

Example 7.

$-2x + 9 \geq 15 \Rightarrow -2x \geq 6 \Rightarrow x \leq -3$.

The solid dot shows that -3 is included in the solution, as are all the values to the "left" of -3.

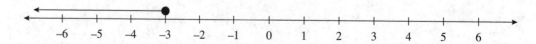

Example 8.

Given that $4x - 5 \leq 3$, what is the greatest possible value of x?

Add 5 to both sides of inequality: $4x \leq 8$

Divide both sides of inequality by 4: $x \leq 2$

Therefore, the greatest possible value of x is 2.

Example 9.

Graph the solutions to the inequality $4x - 2y > -10$.

First isolate the y-variable.

Subtract $4x$ from both sides of inequality: $-2y > -4x - 10$

Divide both sides of inequality by -2: $y < 2x + 5$

Remember that dividing by a negative switches the inequality sign.

To graph $y < 2x + 5$, first create a dashed line at $y = 2x + 5$. Since it is an inequality, the graph must include every point that makes the inequality true. Given that the inequality is a less than (<), shade below the dashed line.

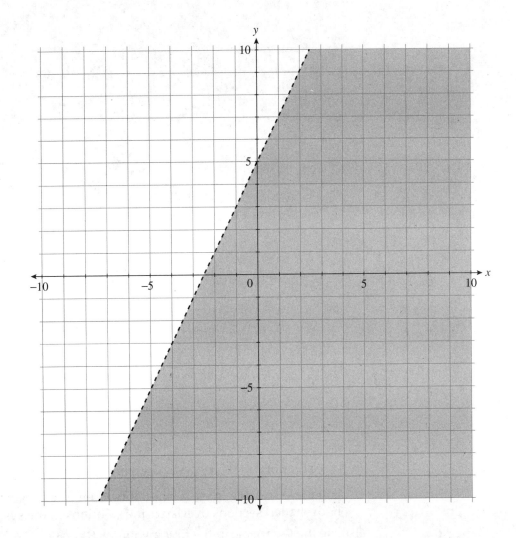

Example 10.

Graph the solutions to the inequality $3x - 6y \le 12$.

First isolate the y-variable.

Subtract $3x$ from both sides of the inequality: $-6y \le -3x + 12$

Divide both sides of the inequality by -6: $y \ge \dfrac{1}{2}x - 2$

Remember that dividing by a negative switches inequality sign.

To graph $y \ge \dfrac{1}{2}x - 2$, first create a solid line at $y = \dfrac{1}{2}x - 2$. Since it is an inequality, the graph must include every point that makes the inequality true. Given that the inequality is a greater than ($\ge$), shade above the line.

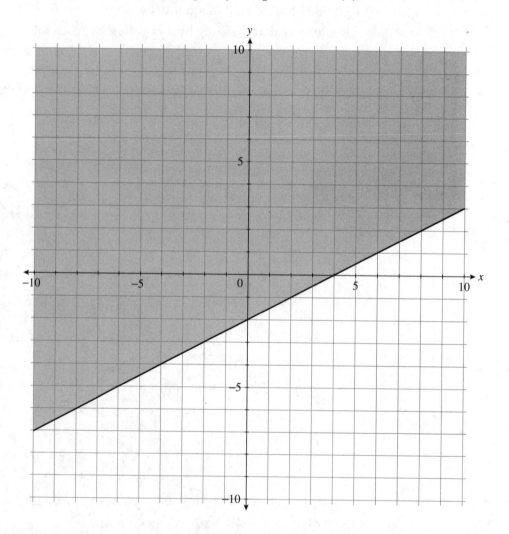

Example 11.

Graph the solutions to the system of inequalities $9x - 3y \ge 12$ and $4x + 2y > 6$.

Isolate the y-variable in each inequality, then graph each. The region where both shaded sections overlap is the solutions to the system of inequalities.

Isolate the y-variable in the first inequality $9x - 3y \ge 12$.

Subtract $9x$ from both sides of the inequality: $-3y \geq -9x + 12$

Divide both sides of the inequality by -3: $y \leq 3x - 4$

Remember that dividing by a negative switches the inequality sign.

To graph $y \leq 3x - 4$, first create a solid line at $y = 3x - 4$. Since it is an inequality, the graph must include every point that makes the inequality true. Given that the inequality is a greater than ($\leq$), shade below the line.

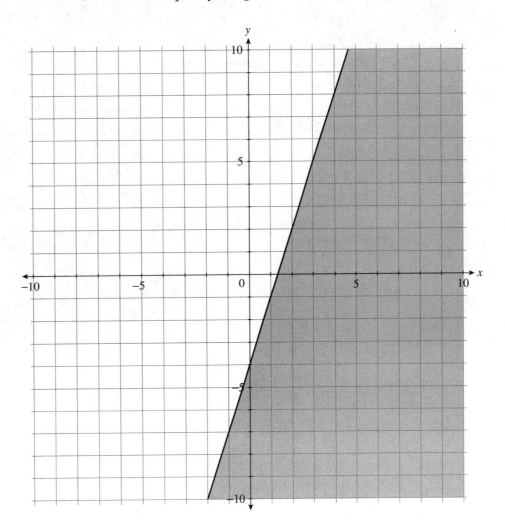

Isolate the y-variable on the second inequality $-2y - 4x < -6$.

Add $4x$ to both sides of inequality: $-2y < 4x - 6$

Divide both sides of inequality by -2: $y > -2x + 3$

Remember that dividing by a negative switches the inequality sign.

To graph $y > -2x + 3$, first create a dashed line at $y = -2x + 3$. Since it is an inequality, the graph must include every point that makes the inequality true. Given that the inequality is a less greater than ($>$), shade above the dashed line. Place this on top of the graph above, and the overlap is the solution to the system of inequalities. This is the dark-shaded region in the graph below.

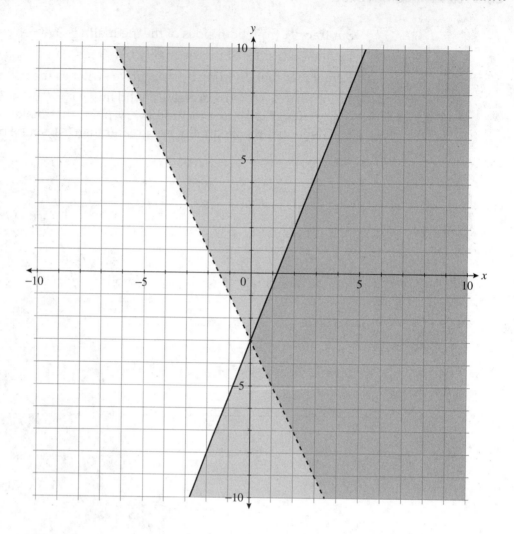

Example 12.

A candle shop makes candles of two different sizes, 4 ounce and 8 ounce. The store sells the 4-ounce candles for $9.99 and the 8-ounce candles for $15.99. On a given day, the store has 400 ounces of wax to use to make the candles and wants to sell the candles for at least a total of $700. Let x represent the number of 4-ounce candles and y represent the number of 8-ounce candles. Assuming the candle store sells every candle made, write a system of inequalities that can be used to determine the possible number of 4-ounce and 8-ounce candles the shop could make.

First write an inequality for the amount of wax that can be used: $4x + 8y \leq 400$

Next write an inequality for the amount of money the shop wants to make: $9.99x + 15.99y \geq 700$

This system of inequalities can be used to determine the possible number of 4-ounce and 8-ounce candles.

Example 13.

A dog walker walks dogs for two different periods of time, 20 minutes and 30 minutes. The dog walker charges $12 for a 20-minute walk and $15 for a 30-minute walk. The dog walker walks dogs for no more than 8 hours a day and wants to earn at least $180 per day. Let x represent the number of 20-minute walks per day and y represent the number of 30-minute walks per day. Write a system of inequalities that can be used to determine the possible number of 20-minute walks and 30-minute walks the dog walker needs to do each day.

First convert 8 hours into minutes: $8(60) = 480$

Next write an inequality for the amount of time spent walking: $20x + 30y \le 480$

Finally write an inequality for the amount of money earned: $12x + 15y \ge 180$

So, the system of inequalities is: $20x + 30y \le 480$

$$12x + 15y \ge 180.$$

Practice Questions

1. Is the point $(2, 3)$ a solution to the inequality $9x - 3y > 8$?

2. Is the point $(-1, -2)$ a solution to the system of inequalities shown below?

$$x - 2y > -2$$

$$2x - 3y \ge 6.$$

3. Graph the solutions to the inequality $4x - 2y \ge 8$.

4. Graph the solutions to the system of inequalities $3y - 9x \le -12$ and $-2y - 4x < -6$.

5. A new internet television provider has two types of subscriptions. The Standard subscription is \$35 per month and the Premium subscription is \$45 per month. Based on past subscription rates, the company averages no more than 115 subscriptions a month, with a revenue of at least \$3,500. Let x represent the number of Standard subscriptions and y represent the number of Premium subscriptions. Write a system of inequalities that describes the possible number of Standard subscriptions, x, and possible number of Premium subscriptions, y, described in the problem.

Practice Answers

1. Substitute $x = 2$ and $y = 3$ into $9x - 3y$ and solve: $9(2) - 3(3) = 18 - 9 = 9$
   Since $9 > 8$, the point $(2, 3)$ is a solution to the inequality $9x - 3y > 8$.

2. Is the point $(-1, -2)$ a solution to the system of inequalities shown below?

$$x - 2y > -2$$

$$2x - 3y \ge 6.$$

Try the first inequality.

Substitute $x = -1$ and $y = -2$ into $x - 2y$ and solve: $(-1) - 2(-2) = -1 + 4 = 3$

Since $3 > -2$, the point $(-1, -2)$ is a solution to the inequality $x - 2y > -2$.

Try the second inequality.

Substitute $x = -1$ and $y = -2$ into $2x - 3y$ and solve: $2(-1) - 3(-2) = -2 + 6 = 4$

Since $4 < 6$, the point $(-1, -2)$ is not a solution to the inequality $2x - 3y \ge 6$.

Since the point $(-1, -2)$ does not work for both inequalities, it is not a solution to the system of inequalities.

3.

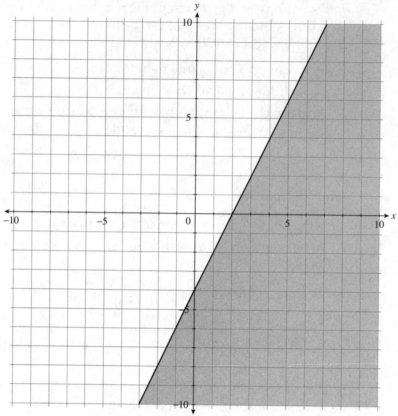

4. The solution is the dark-shaded region in the graph below.

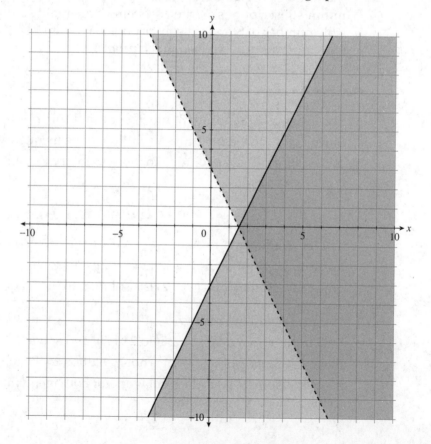

5. $x + y \leq 115$

$35x + 45y \geq 3{,}500.$

SOLVED SAT PROBLEMS

1. A parking garage has two different rates, $15 for standard-size vehicles and $20 for oversized vehicles. The garage can fit a maximum of 200 vehicles including not more than 150 oversized vehicles. Each day the garage has a goal of earning at least $2,000. Which of the following systems of inequalities represents the conditions described if x is the number of standard-size vehicles and y is the number of oversized vehicles?

A. $15x + 20y \leq 2{,}000$
   $x + y \geq 200$
   $y \geq 150$

B. $15x + 20y \geq 2{,}000$
   $x + y \leq 200$
   $y \leq 150$

C. $15x + 20y \leq 2{,}000$
   $x + y \leq 200$
   $y \leq 150$

D. $15x + 20y \geq 2{,}000$
   $x + y \geq 200$
   $y \geq 150$

EXPLANATION: Choice B is correct.

The problem states that the garage cannot fit more than 150 oversized vehicles. This corresponds to the inequality $y \leq 150$. Therefore, choice A and choice D can be eliminated. We also know that the total number of vehicles cannot exceed 200. This corresponds to the inequality $x + y \leq 200$, found in both of the remaining choices, B and C. The final piece of information is that the garage has a goal of earning at least $2,000, which corresponds to the inequality $15x + 20y \geq 2{,}000$. Of the choices remaining, this is found only in choice B.

2. Which of the following ordered pairs (x, y) satisfies the system of inequalities shown below?

$$x - 2y \leq 4$$
$$2x - y < 3$$

A. $(-5, -6)$
B. $(-2, -4)$
C. $(2, 4)$
D. $(5, 6)$

EXPLANATION: Choice C is correct.

Try plugging each choice into the inequalities to eliminate incorrect choices.

Try $(-5, -6)$ in the first inequality.

Substitute $x = -5$ and $y = -6$ into $x - 2y$ and solve:
  $(-5) - 2(-6) = -5 + 12 = 7$

Since $7 > 4$, the $(-3, -6)$ is not a solution to the inequality
  $x - 2y \leq 4$.

Try $(-2, -4)$ in the first inequality.

Substitute $x = -2$ and $y = -4$ into $x - 2y$ and solve:
  $(-2) - 2(-4) = -2 + 8 =$

Since $6 > 4$, $(-2) - 2(-4) = -2 + 8 = 6$ is not a solution to the inequality
  $x - 2y \leq 4$.

Try $(2, 4)$ in the first inequality.

Substitute $x = 2$ and $y = 4$ into $x - 2y$ and solve: $(2) - 2(4) = 2 - 8 = -6$

Since $-6 < 4$, $(2, 4)$ is a solution to the inequality $x - 2y \leq 4$.

Try $(2, 4)$ in the second inequality.

Substitute $x = 2$ and $y = 4$ into $2x - y$ and solve: $2(2) - (4) = 4 - 4 = 0$

Since $0 < 3$, $(2, 4)$ is a solution to the inequality $2x - y < 3$.

So $(2, 4)$ is a solution to both inequalities; therefore $(2, 4)$ is a solution to the system of inequalities.

INEQUALITIES
PRACTICE SAT QUESTIONS

ANSWER SHEET

Choose the correct answer.

1. Ⓐ Ⓑ Ⓒ Ⓓ	11. Ⓐ Ⓑ Ⓒ Ⓓ
2. Ⓐ Ⓑ Ⓒ Ⓓ	12. Ⓐ Ⓑ Ⓒ Ⓓ
3. Ⓐ Ⓑ Ⓒ Ⓓ	13. Ⓐ Ⓑ Ⓒ Ⓓ
4. Ⓐ Ⓑ Ⓒ Ⓓ	14. Ⓐ Ⓑ Ⓒ Ⓓ
5. Ⓐ Ⓑ Ⓒ Ⓓ	
6. Ⓐ Ⓑ Ⓒ Ⓓ	
7. Ⓐ Ⓑ Ⓒ Ⓓ	
8. Ⓐ Ⓑ Ⓒ Ⓓ	
9. Ⓐ Ⓑ Ⓒ Ⓓ	
10. Ⓐ Ⓑ Ⓒ Ⓓ	

PRACTICE SAT QUESTIONS

1. In the inequality $5r - 3 \leq 1$, what is the greatest possible value of $5r + 1$?
    A. 3
    B. 4
    C. 5
    D. 6

2. Which of the following values could be substituted for x and y in the inequality $-2x + 3y \geq 10$?
    I. 4. 6
    II. 2, 3
    III. $-3, 2$
    A. I only
    B. II only
    C. I and III only
    D. II and III only

3. $y > 3x - 2$

    $3x > 5$

    Which of the following is a y that is a valid answer to this system of equations?

    A. $y > \dfrac{2}{3}$

    B. $y > \dfrac{5}{3}$

    C. $y > 0$

    D. $y > 3$

4. Eric has two part-time jobs to make ends meet. He is a salesperson making \$14 an hour plus \$10 commission on each sale, and works as a tutor, which pays \$25 an hour but with a tip every fourth session of \$15. He makes no other money and has no other jobs. He cannot work any more than 15 hours a week due to his family responsibilities, and his budget indicates that he must earn at least \$320 per week not counting any commissions or tips he receives. Which of the following systems of inequalities represents Eric's situation in terms of x and y? Assume that x is the number of hours he works in his sales job and y is the number of hours Eric works as a tutor in any given week.

    A. $\begin{cases} 14x + 25y \leq 320 \\ x + y \geq 15 \end{cases}$

    B. $\begin{cases} 14x + 25y \leq 320 \\ x + y \leq 15 \end{cases}$

    C. $\begin{cases} 14x + 25y \geq 320 \\ x + y \leq 15 \end{cases}$

    D. $\begin{cases} 14x + 25y \geq 320 \\ x + y \geq 15 \end{cases}$

5. $y \geq 2x + 2$

    $-x < 5 - y$

    Which of the ordered pairs is a solution to both these inequalities?

    A. $(2, 6)$
    B. $(6, 2)$
    C. $(6, 9)$
    D. $(9, 6)$

6. Which of the following ordered pairs (x, y) satisfies the inequality $10x - 6y < 8$?
    I. $(2, 2)$
    II. $(4, 10)$
    III. $(6, 4)$
    A. I only
    B. II only
    C. I and II only
    D. II and III only

7. There were 20 cabins on the Mississippi paddle wheeler *Mark Twain*. Single (s) cabins were $135, while it cost $220 for a double ($d$). On a recent trip, some or all of the 20 cabins were sold and passengers paid at most $3,720 in total for the cabins. Which of the following systems of inequalities best matches cabin sales on that trip?

A. $s + d \geq 20$
$\$220s + \$135d \geq 3,720$

B. $s + d \geq 20$
$\$135s + \$220d \leq 3,720$

C. $s + d \leq 20$
$\$135s + \$220d \leq 3,720$

D. $s + d \leq 20$
$\$220s + \$135d \geq 3,720$

8. A baker is buying flour and sugar, but can get no more than 400 pounds delivered at a time in a single shipment. Each sack of flour weighs 30.4 pounds, and each bag of sugar weighs 11.33 pounds. He also must buy 20 pounds of chocolate, which needs to be included in the shipment. The bakery's owner decides they need to buy at least three times as many containers of sugar as containers of flour. f represents the number of containers of flour, and s denotes the number of containers of sugar. Which of the following systems represents the baker's scenario?

A. $\begin{cases} 11.33s + 30.4f \leq 380 \\ s \geq 3f \end{cases}$

B. $\begin{cases} 11.33s + 30.4f \leq 380 \\ 3s \geq f \end{cases}$

C. $\begin{cases} 33.99s + 30.4f \leq 400 \\ s \geq 3f \end{cases}$

D. $\begin{cases} 33.99s + 30.4f \leq 400 \\ 3s \geq f \end{cases}$

9. $x + y > 9$ and $3x - 2y > 20$

Which ordered pair (x, y) satisfies the system of equations shown?

A. $(-6, -4)$
B. $(-6, 4)$
C. $(6, 4)$
D. $(12, 2)$

10. $y + x < p$
$y > q + x$

If $(2, 2)$ is a solution to this system of equations, which of the following best characterizes the relation between p and q?

A. $p = q$
B. $p < q$
C. $p > q$
D. $-p > q$

11. The mountain climbing team had two types of ropes that could be tied together. One type of rope, x, cost $4.00 a linear foot. Another type, y, cost $7.50 a linear foot. The combined ropes have to be at least 100 feet long and cost less than $600. Select the system of equations below that best models this situation.

A. $x + y \geq 100$
$\$4.00 x + \$7.50 y \geq \$600$

B. $x + y \geq 100$
$\$4.00 x + \$7.50 y \leq \$600$

C. $x + y \leq 100$
$\$4.00 x + \$7.50 y \geq \$600$

D. $x + y \geq 100$
$\$4.00 x + \$7.50 y < \$600$

12.

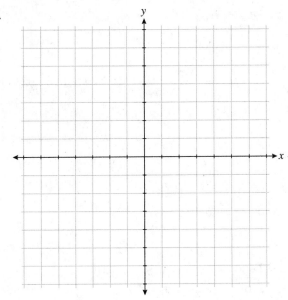

The chart above shows the four unlabeled quadrants on the coordinate plane.

$y \leq -2x + 3$

$y \geq 3x + 2$

Which quadrant on the coordinate plane contains no solution for the system of linear inequalities shown above?

A. Quadrant I
B. Quadrant II
C. Quadrant III
D. Quadrant IV

13. $12x > 4y + 8$

Which of the following inequalities is equivalent to the inequality above?

A. $3x - y > 2$
B. $4y + 8 > 12x$
C. $4y + 8 < -12x$
D. $3x + y > 2$

14. $2x - 1 \leq 6x + 11$

Which of the following choices could not be a solution to the inequality above?

A. -10
B. 0
C. 5
D. 10

EXPLAINED ANSWERS

1. **EXPLANATION: Choice C is correct.**

   $5r - 3 \leq 1$. To get to $5r + 1$ from $5r - 3$, we need to add 4, because $1 - (-3)$ is 4. Add 4 to both sides to get $5r + 1 \leq 5$.

2. **EXPLANATION: Choice C is correct.**

   Use trial and error to solve this problem.

   I Substitute (4, 6) $-2x + 3y \geq 10$ $-2(4) + 3(6) = -8 + 18 = 10$ True

   II Substitute (2, 3) $-2x + 3y \geq 10$ $-2(2) + 3(3) = -4 + 9 = 5$ False

   III Substitute (−3, 2) $-2x + 3y \geq 10$ $-2(-3) + 3(2) = 6 + 6 = 12$ True.

3. **EXPLANATION: Choice D is correct.**

   You can substitute (5) for $3x$ in the top inequality. $y > 5 - 2$ and simplify ($y > 3$).

   You can also solve this like a system of equations. Stack the inequalities and add.

   $$y > 3x - 2$$
   $$0 > -3x + 5$$
   $$\overline{y > 3} \quad \longrightarrow \quad \text{Choice D.}$$

4. **EXPLANATION: Choice C is correct.**

   Use one set of information at a time, as well as process of elimination, to help you solve this problem. The two types of hours need to add up to no more than 15 hours a week according to the problem, so they need to be less than or equal to 15: eliminate A and D. Now, deal with the other equation: the amount of money he makes must be $320 or more, so that leaves choice C as the only option.

5. **EXPLANATION: Choice A is correct.**

   Plug the answer choices in to the inequalities one by one to see which might work. If you got something other than answer choice A, you may have made a sign mistake: the SAT often tests sign rules with inequalities, so be careful!

6. **EXPLANATION: Choice B is correct.**

   Plug the numbers carefully into the inequality: Try I first. $10(2) - 6(2) < 8$ or $20 - 12 < 8$. Since $8 = 8$, this is false. Eliminate choices A and C. Now try III: $10(6) - 6(14) < 8$, or $60 - 40 < 8$. Since $20 > 8$, this is false. Note that you don't have to try II, because it appears in both of the remaining answer choices after you've eliminated I.

7. **EXPLANATION: Choice C is correct.**

   Use process of elimination and deal with one set of equations at a time. The total number of cabins can't exceed 20: that gets rid of choices A and B. The $220 goes with the d variable, so C is the only possible answer.

8. **EXPLANATION: Choice B is correct.**

   Sugar needs to be three times as many bags as flour, so $3s$ must be greater than or equal to f, which gets rid of choices A and C. In choice D, the numbers in the first equation don't reflect the correct values from the problem. Don't forget that the total needs to be 380 to account for the chocolate, not 400.

9. **EXPLANATION: Choice D is correct.**

   Only choices C and D satisfy the first of the two inequalities. After that, plug the numbers in to see which pair works for the second inequality as well.

10. **EXPLANATION: Choice C is correct.**

    Insert the solution $(2, 2)$ into the system of equations to find the relationship: $2 + 2 < p$, $4 < p$. Then $y - x > q$, $0 > q$. $p > 4$ and $q < 0$. Therefore, $p > q$.

11. **EXPLANATION: Choice D is correct.**

    The two types of ropes must be greater than or equal to 100 feet in total, so that's a greater-than-or-equal-to equation for the first one: eliminate C. For the other equation, the total price must be less than, not less than or equal to, $600. Careful! Not every set of answer choices for these problems is the same, and there can be traps in small things like the "equal to" part of the equation.

12. **EXPLANATION: Choice D is correct.**

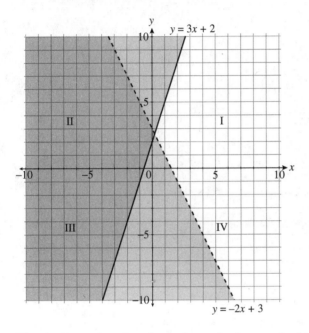

    The dark overlapping shaded area in the graph above shows all the solutions to the system of equations. The dark shading is evident in Quadrants II and III. The idea is to see that dark shading just creeps over into Quadrant I. Quadrant IV does not contain the dark overlapping shaded regions, so it is the answer to the question.

13. **EXPLANATION: Choice A is correct.**

    Divide all parts of the inequality by 4 to get $3x > y + 2$, and then subtract y from both sides to get something in the form of one of the answer choices, A. If you picked another choice, you may need to work on what to do with negative signs for inequality problems.

14. **EXPLANATION: Choice A is correct.**

    You have a choice between plugging these values into the equation and simply solving. If you choose to plug in, you'll find that $2(-10) - 1 = -21$, which is not less than or equal to $6(-10) + 11$ (which equals -49). To solve the equation instead, subtract $6x$ from both sides to get $-4x - 1 \leq 11$, and then add the 1 to both sides: $-4x \leq 12$. Divide both sides by -4 to get $x \geq -3$: the only choice that doesn't fit, then, is A, which is less than -3.

CHAPTER 10

ABSOLUTE VALUE

The **absolute value** of an expression is best thought of as the distance a value is from zero; therefore, the smallest value that can be obtained by taking the absolute value is 0, absolute value of 0, $|0| = 0$. Otherwise the result will be positive: the absolute value of 2, $|2| = 2$ and the absolute value of -2, $|-2| = 2$. Problems on the SAT will ask you to solve equation, inequalities and problems that further develop these ideas. Begin with the mathematics review and then complete and correct the practice problems. There are 2 Solved SAT Problems and 13 Practice SAT Questions with answer explanations.

ABSOLUTE VALUE

Example 1.

Solve equation for x. $|x| = 3$

The shortcut is to write two equations that look like this,

$x = 3$ or $x = -3$

SOLUTION: x can be 3 or -3.

Example 2.

Solve the equation for x.

$|x - 4| = 6$

Write two equations.

In one, the quantity inside the absolute value equals the quantity on the other side of the equation.

In the other, the quantity inside the absolute value equals the negative ($-$) of the quantity on the other side of the equation.

Solve both the equations.

$$x - 4 = 6 \implies x = 10$$
$$|x - 4| = 6 \implies \quad \text{or}$$
$$x - 4 = -6 \implies x = -2$$

SOLUTION: $x = 10$ or $x = -2$.

Let's check.

Substitute 10 for x: $|x - 4| = 6 \implies |10 - 4| = 6 \implies |6| = 6$. It checks

Substitute -2 for x: $|x - 4| = 6 \implies |-2 - 4| = 6 \implies |-6| = 6$. It checks.

Remember an absolute value is always positive.

Example 3.

Solve the inequality for *x*.

$|x + 11| = 0$
$$x + 11 = 0 \implies x = -11$$
$|x + 11| = 0$
$$x + 11 = 0 \implies x = -11$$

The equations are the same, so there is just one solution.

Example 4.

Solve the inequality for x

$$|x| + 31 \leq 18$$

Simplify. $|x| + 31 \leq 18 \implies |x| \leq -13$

That can't be. The absolute value of *x* must be positive. However, this inequality says $|x|$ is negative. There is no solution to this problem.

Example 5.

Solve the inequality for *x*.

$|x - 2| \leq 7$
$$x - 2 \leq 7 \implies x \leq 9$$
$|x - 2| \leq 7$
$$x - 2 \geq -7 \implies x \geq -5$$

SOLUTION: $x \geq -5$ and $x \leq 9$

Write it this way. $-5 \leq x \leq 9$.   The value *x* is from −5 through 9.

The solid dots show that −5 and 9 are included.

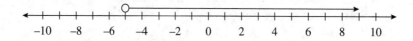

Example 6.

Solve inequality for *x* and graph the solution on the number line.

Simplify the inequality. $|2x + 7| - 6 > 11 \implies |2x + 7| > 17$

Then write two inequalities, just like we did with equations, and solve.

$$2x + 7 > 17 \implies 2x > 10 \implies x > 5$$
$$\implies \qquad\qquad \text{or}$$
$$2x + 7 < -17 \implies 2x < -24 \implies x < -12.$$

Practice Problems

1. If $|4x - 12| + 15 = 19$, what are the possible values of x?
2. If $|2x + 7| \leq 13$, what are the possible values of x?
3. If $|x + 15|$, what are the possible values of x?
4. If $|y - 10| = 0$, what are the possible values of y?
5. If $|y - 1| = 2y + 3$.

Practice Answers

1. $x = 4$ or $x = 2$.
2. $-10 \leq x \leq 3$.
3. No solution.
4. $y = 10$.
5. $y = -\dfrac{2}{3}$, ($y = -4$ cannot be a solution because does not check as correct.)

SOLVED SAT PROBLEMS

1. Which of the following expressions is equal to 2 for some value of x?

   A. $3 + |x - 2|$

   B. $3 + |2 - x|$

   C. $3 + |x - 2|$

   D. $3 - |2 - x|$

   EXPLANATION: Choice D is correct.

   The absolute value of an expression is never less than 0, so adding an absolute value to 3, will never be less than 3. Therefore, to get an answer of 2, you must subtract the absolute value from 3, which is choice D.

2. A point with a coordinate of x on a number line is 5 units away from 7. Which of the following equations can be used to find a number x?

A. $|x - 5| = 7$

B. $|x + 5| = 7$

C. $|x - 7| = 5$

D. $|x + 7| = 5$

EXPLANATION: Choice C is correct.

The distance between two points on a number line is a positive number. To solve this problem, set the absolute value of the difference between the points, x and 7, equal to the distance 5. That is, find the absolute value of the difference between x and 7 and set it equal to 5.

Check your answer. Five units to the left of 7 is 2. Five units to the right is 12.

$$x - 7 = 5 \implies x = 12$$

$$|x - 7| = 5$$

$$x - 7 = -5 \implies x = 2$$

It checks. Notice that $|7 - x|$ also works.

ABSOLUTE VALUE
PRACTICE SAT QUESTIONS

ANSWER SHEET

Choose the correct answer.
If no choices are given, grid the answers in the section at the bottom of the page.

1. Ⓐ Ⓑ Ⓒ Ⓓ
2. Ⓐ Ⓑ Ⓒ Ⓓ
3. Ⓐ Ⓑ Ⓒ Ⓓ
4. Ⓐ Ⓑ Ⓒ Ⓓ
5. Ⓐ Ⓑ Ⓒ Ⓓ
6. Ⓐ Ⓑ Ⓒ Ⓓ
7. Ⓐ Ⓑ Ⓒ Ⓓ
8. **GRID**
9. Ⓐ Ⓑ Ⓒ Ⓓ
10. Ⓐ Ⓑ Ⓒ Ⓓ

11. Ⓐ Ⓑ Ⓒ Ⓓ

Use the answer spaces in the grids below if the question requires a grid-in response.

| Student-Produced Responses | ONLY ANSWERS ENTERED IN THE CIRCLES IN EACH GRID WILL BE SCORED. YOU WILL NOT RECEIVE CREDIT FOR ANYTHING WRITTEN IN THE BOXES ABOVE THE CIRCLES. |

8.

	/	/	/
	.	.	.
.	0	0	0
1	1	1	1
2	2	2	2
3	3	3	3
4	4	4	4
5	5	5	5
6	6	6	6
7	7	7	7
8	8	8	8
9	9	9	9

PRACTICE SAT QUESTIONS

1. If $|x - y| = d$ and $x > y$, which of the following cannot be the value of $y - x$?

   I. d
   II. $-d$
   III. $|y - x|$

   A. I
   B. II
   C. I and II
   D. I and III

2. $a = ||x|^3 - |x - y|^2 - y + x|$. If $x = -2$ and $y = 5$, $a =$

   A. 47
   B. 48
   C. 49
   D. 50

3. Which of the following inequalities create the graph below?

   A. $|x - 8| < 4$
   B. $|-x - 3| < 8$
   C. $|x - 4| < 8$
   D. $|-x - 4| > 8$

   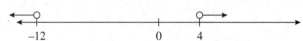

   -12 0 4

4. Which of the following values of x creates a result of -2 in the expression below?

   $$|x - 3| + 2$$

   A. 4
   B. 5
   C. 6
   D. There is no such value

5. Which expression below is equal to 0 when x is 0?

   A. $|x - 2| - 2$
   B. $|x + 2| + 2$
   C. $|2 - x| + 2$
   D. $|2 + x| + 2$

6. Which of these choices is a solution of the given equation?

   $$4|x + 3| = 7|x + 3|$$

   A. -3
   B. 0
   C. 3
   D. There are no solutions

7. What are the solutions of the following equation?

   $$|x + 10| + 12 = 30$$

   A. $-8, -28$
   B. $8, -28$
   C. $28, 8$
   D. $28, -8$

8. Solve for s in the equation below

   $$4|3 - 3s| - 18 = 66, s < 0$$

9. For what value of x is $|x - 2| + 2$ equal to 0?

   A. -2
   B. 0
   C. 1
   D. There is no such value of x

10. For at least one value of p, which of the following expressions is equivalent to 0?

    A. $|p - 2| - 2$
    B. $|1 - p| + 1$
    C. $|p + 1| + 3$
    D. $|p - 1| + 1$

11. $|x^2 + 4| = 8$

    Which of the following is a real number solution to the equation above?

    A. 4
    B. -2
    C. -4
    D. -8

EXPLAINED ANSWERS

1. **EXPLANATION: Choice D is correct.**

   We know $x > y$ so $y - x$ is negative.
   $$|x - y| = d \quad y - x = -d$$

   That means $(y - x)$ cannot equal these choices.

   I. d

   III. $|y - x|$.

2. **EXPLANATION: Choice B is correct.**

   Substitute and solve for a.

   $$a = \left||-2|^3 - |-2 - 5|^2 - 5 + (-2)\right| = \left|2^3 - 7^2 - 5 - 2\right|$$
   $$|8 - 49 - 5 - 2| = |-48| = 48.$$

3. **EXPLANATION: Choice D is correct.**

   The arrows are pointing out. Choose a greater-than inequality. The midpoint of -12 and 4 is -4. This is only true for choice D. Check your solution.

   $$-x - 4 > 8 \Rightarrow -x > 12 \Rightarrow x < -12.$$
   $$|-x - 4| > 8$$
   $$-x - 4 < -8 \Rightarrow -x < -4 \Rightarrow x > 4.$$

   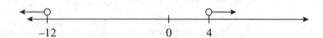

4. **EXPLANATION: Choice D is correct.**

   Solve the equation by creating an equation: $|x - 3| + 2 = -2$. Subtract the 2 to the right side to yield: $|x - 3| = -4$. There is no value that results from an absolute value that will result in a negative number. Therefore, the answer is D. Choices A, B, and C, when inserted into the equation, will not also not produce a negative number and therefore will not yield a result of -4.

5. **EXPLANATION: Choice A is correct.**

   Plug in 0 where you see x. Inside the absolute value expression in A, plugging in 0 gives us -2, which has an absolute value of 2. Subtract 2 to equal 0, as required by the question. If you selected another choice, you likely made a math error to do with signs. All other answer choices yield an expression greater than 2.

6. **EXPLANATION: Choice A is correct.**

   Plug in the answer choices. Only A makes both sides of the equation equal.

7. **EXPLANATION: Choice B is correct.**

   In order to solve for both answers to the equation, start by simply removing the absolute value from the equation: $x + 10 + 12 = 30$, so $x = 8$. Two answers contain that number, but to solve for the other value, simply distribute a negative sign in front of the absolute value: $-|x + 10| + 12 = 30$, which becomes $-x - 10 + 12 = 30$. Simplify: $-x + 2 = 30$, $-x = 28$, $x = -28$.

   You can also simplify first: $|x + 10| = 18$. Then, $x + 10 = 18$ or $x + 10 = -18$. Solving gives you $x = 8$ and $x = -28$.

8. **EXPLANATION: The correct answer is −6.**

   Solve the equation without the absolute value first: $4|3 - 3s| - 18 = 66$, so $4|3 - 3s| = 84$. Divide by 4 on each side to yield: $|3 - 3s| = 21$. Simplify to $3 - 3s = \pm 21$. Then subtract 3 from both sides to yield $-3s = -24$ or 18. Divide by -3 to solve for s: $s = 8$ or -6. Finally, the question imposed the condition that $s < 0$, so -6 must be the solution. Therefore, $s = -6$. The SAT will ask you to grid in the answer without the negative sign, if a negative answer is indicated.

9. **EXPLANATION: Choice D is correct.**

   This question tests your understanding of absolute value as distance from zero. To make the problem work, there would have to be a negative result to the absolute value part of the equation: it would have to equal -2. That's impossible, so there's no solution for this question.

10. **EXPLANATION: Choice A is correct.**

    For the expression in A to be equal to 0, the amount in the absolute value must equal 2: this is easy to do by making $p = 4$. For B, it would need to equal -1: similarly, impossible. For C, however, the amount in the absolute value would need to equal -3, which is impossible. For D, the same is true. Thus, A is the only possible answer.

11. **EXPLANATION: Choice B is correct.**

    The expression tells you that the absolute value is 8, meaning that either $x^2 + 4 = 8$ or $x^2 + 4 = -8$. Using the second of those two equations will have you taking the square root of a negative, which would not be a real number, so focus on $x^2 + 4 = 8$. Subtract 4 from both sides: $x^2 = 4$, which means $x = \pm 2$. Only one of those values is in an answer choice.

CHAPTER 11

REWRITING EQUATIONS AND EXPRESSIONS

This section reviews the fundamental algebra skills of multiplying, simplifying, factoring, and problem-solving that you need to succeed on the SAT. Some SAT problems simply ask you to apply this knowledge in a straightforward way. Other times these skills are only a part of what is necessary to solve the problem. Begin with the mathematics review and then complete and correct the practice problems. There are 2 Solved SAT Problems and 16 Practice SAT Questions with answer explanations.

EXPRESSIONS

Expressions are numbers, variables, and operations grouped together to represent a value. Expressions do not have equal signs and are not equations, All Polynomials are expressions. Not all expressions are Polynomials. The rules for working with polynomials are the same as the rules for working with expressions.

POLYNOMIALS

A Polynomial is the sum of a finite number of terms, in which all variables have whole number exponents and no variables are in a denominator.

Each term in a polynomial has the form ax^n where **a** is the coefficient, **x** is the variable and **n** is a whole number exponent. A few examples will help clarify things.

These Are Polynomials

5 (5 can be written as $5x^0$)

$\frac{1}{3}x$, $-2.6x^2$, (Fractions, decimals and even numbers like π, are OK as coefficients.)

$3x + 4y^0 - 6$, (0 is a whole number so it is OK as an exponent)

$3y + 6y^2 - zy^3 - 5$ (The definition says sum. Think of this polynomial as: $3y + 6y^2 + (-zy^3) + (-5)$.

These Are Not Polynomials

$5z^{1/3}$ (No fractions as exponents)

$7y^{-3}$ (No negative numbers as exponents)

$\frac{3}{y} + 2$ (No variables in a denominator).

EQUATIONS

An equation says that two quantities are equal. Equations have equal signs. An equation always has two equal expressions. Sometimes one or more of those expressions is a polynomial.

SIMPLIFYING POLYNOMIALS

Multiply each term in one polynomial to each term in the other polynomial. Finally, add and subtract like terms to complete the process.

Example 1.

Simplify: $(4x^2 + 5x - 3)(x + 5) - (2x + 7)(x - 4) =$

Multiply Terms. $(4x^3 + 20x^2 + 5x^2 + 25x - 3x - 15) - (2x^2 - 8x + 7x - 28) =$

Combine like terms in parentheses.

$(4x^3 + 25x^2 + 22x - 15) - (2x^2 - x - 28) =$

Distribute the negative.

$4x^3 + 25x^2 + 22x - 15 - 2x^2 - (-x) - (-28) = 4x^3 + 25x^2 + 22x - 15 - 2x^2 + x + 28$

Group like terms.

$4x^3 + (25x^2 - 2x^2) + (22x + x) + (-15 + 28) =$

Combine like terms.

$4x^3 + 23x^2 + 23x + 13$

FACTORING POLYNOMIALS

When factoring polynomials, check first to see if there is a common factor to all the terms. When factoring is complete, check the answer by multiplying the polynomial.

Binomials

You will multiply two binomials and simplify (factor) a trinomial as the product of two binomials.

Multiplying Binomials

Use the FOIL Method to Multiply Binomials.

Here is an overview and an example.

$(ax + b)(cx + d) = (ax \times cx) + (ax \times d) + (cx \times b) + (b \times d) = (ac)x^2 + \underline{(ad)x + (cb)x} + (bd)$

Find the solution.

$(3x + 4)(5x + 6) =$

Use FOIL $(3x + 4)(5x + 6) = (3x \times 5x) + (3x \times 6) + (5x \times 4) + (4 \times 6) =$

$\qquad\qquad\qquad\qquad\qquad\qquad$ **F** $\qquad$ **O** $\qquad$ **I** $\qquad$ **L**

Multiply. $\qquad\qquad\qquad 15^2 + 18x + 20x + 24$

Combine Like Terms. $\quad 15^2 + 38x + 24$

Example 2.

Simplify $x^2 - 13x + 42 =$

Rewrite the middle term $-13x$ with $\qquad\qquad x^2 - \underline{6x} - \underline{7x} + 42$
coefficients whose product is 42. Notice
that $-6 \times -7 = 42$. That's what we want.

Use parentheses to form two binomials. $\qquad (x^2 - 6x) + (-7x + 42)$

Factor out the common factor x from the
first binomial, and factor out the common $\qquad x(x - 6) - 7(x - 6)$
factor -7 from the second binomial.

Factor out the common factor $(x - 6)$. $\qquad (x - 6)(x - 7).$

$x^2 - 13x + 42 = (x - 6)(x - 7).$

Example 3.

Simplify $6x^2 + 22x + 12 =$

Factor out 2 to simplify the trinomial. $\qquad 2(3x^2 + 11x + 6) =$
The trinomial inside the parentheses is in
simplest form. Solve that trinomial.

The trinomial $3x^2 + 11x + 6$ is in descending order. The leading coefficient is 3 ($\underline{3}x^2$).
Use this approach when the leading coefficient is not 1.

Find the factors of the first term. $\qquad\qquad 3x^2 = (3x)(x)$

Use the factors to begin to form a binomial pair. $\quad (3x + \underline{\quad})(x + \underline{\quad})$

Find the factors of the last term. $\qquad\qquad 6 = (6 \times 1)$ and $6 = (2 \times 3)$

Substitute the factors in the binomial pair $\qquad$ That's $(3x + \underline{2})(x + \underline{3})$
that make the middle term $11x$.

It checks. The middle term is correct. $\qquad 9x + 2x = \underline{11x}$

Remember to multiply by 2. $\qquad\qquad 2(3x^2 + 11x + 6) = 2(3x + \underline{2})(x + \underline{3})$

Example 4.

The formula for surface area, S, of a cylinder in terms of the radius, r, and the height, h, is $S = 2\pi rh + 2\pi r^2$. Write an equation for h in terms of S and r.

Subtract $2\pi r^2$ from both sides of equation: $S - 2\pi r^2 = 2\pi rh$

Divide both sides of equation by $2\pi r$: $\dfrac{S - 2\pi r^2}{2\pi r} = h$

Therefore, $h = \dfrac{S - 2\pi r^2}{2\pi r}$ or $\dfrac{S}{2\pi r} - r$.

Difference of Squares

The difference of squares follows a standard form.

$(a^2 - b^2) = (a - b)(a + b)$

Example 5.

Simplify $3x^2 - 27 = 3(x^2 - 9) = 3(x - 3)(x + 3)$.

Square of a Binomial

The square of a binomial follows a standard form.

$(a + b)^2 = a^2 + 2ab + b^2$

Example 6.

Simplify $(8x^2 + 24xy + 18y^2) = 2(4x^2 + 12xy + 9y^2) =$

$(4x^2 + 12xy + 9y^2)$ is the $(2x + 3y)^2$

$2(2x + 3y)^2$

Practice Problems

1. Simplify $(-3x^2 + 7x + 5)(x - 2) + (3x - 4)(x + 8)$.
2. Factor $27x^2 - 72xy + 48y^2$.
3. Factor $50x^2 - 72$.
4. The formula for surface area, S, of a sphere in terms of the radius, r, of the sphere is $S = 4\pi r^2$. Write an equation for r in terms of S.
5. Factor the expression: $4s^4 - 4t^4$.
6. $y = \dfrac{x - d}{z}$. Solve the equation for x.

Practice Answers

1. $-3x^3 + 16x^2 + 11x - 42$.

2. $3(3x - 4y)(3x - 4y) = 3(3x - 4y)^2$.

3. $2(5x + 6)(5x - 6)$.

4. $r = \sqrt{\dfrac{S}{4\pi}}$.

5. $4(s^2 - t^2)(s^2 + t^2)$. Note $s^2 - t^2$ can be further factored by difference of squares into $(s - t) \times (s + t)$.

6. $x = zy + d$.

SOLVED SAT PROBLEMS

1. Which of the following is equivalent to $\left(\dfrac{x}{4} - 2y\right)^2$?

A. $\dfrac{x^2}{16} - 4y^2$

B. $\dfrac{x^2}{16} + 4y^2$

C. $\dfrac{x^2}{16} - \dfrac{xy}{2} + 4y^2$

D. $\dfrac{x^2}{16} - xy + 4y^2$

EXPLANATION: Choice D is correct.

Rewrite $\left(\dfrac{x}{4} - 2y\right)^2$: $\left(\dfrac{x}{4} - 2y\right)\left(\dfrac{x}{4} - 2y\right)$

Distribute: $\left(\dfrac{x}{4}\right)\left(\dfrac{x}{4}\right) - \left(\dfrac{x}{4}\right)2y - \left(\dfrac{x}{4}\right)2y + (2y)(2y)$

Multiply: $\dfrac{x^2}{16} - \dfrac{xy}{2} - \dfrac{xy}{2} + 4y^2$

Combine like terms: $\dfrac{x^2}{16} - \dfrac{2xy}{2} + 4y^2 = \dfrac{x^2}{16} - xy + 4y^2$.

2. Which of the following is equivalent to $2x^2 - 4y^2$?

    A. $(\sqrt{2}x - 2y^2)(\sqrt{2}x + 2y^2)$

    B. $(\sqrt{2}x - 2y^2)(\sqrt{2}x - 2y^2)$

    C. $(\sqrt{2}x - 2y)(\sqrt{2}x + 2y)$

    D. $(\sqrt{2}x - 2y)(\sqrt{2}x - 2y)$

EXPLANATION: Choice C is correct.

Looking at the original question and the answers, notice that they are asking to factor the given equation as the difference of two squares.

$$\sqrt{2x^2} = \sqrt{2}x \text{ and } \sqrt{4y^2} = 2y$$

Using the rules about the difference of two squares gives

$$2x^2 - 4y^2 = (\sqrt{2}x - 2y)(\sqrt{2}x + 2y).$$

Don't be afraid to FOIL out your answer choices if necessary. If you do this, start with the answer choice that you think is most likely correct.

REWRITING EQUATIONS AND EXPRESSIONS
PRACTICE SAT QUESTIONS

ANSWER SHEET

Choose the correct answer.
If no choices are given, grid the answers in the section at the bottom of the page.

1. (A) (B) (C) (D)
2. (A) (B) (C) (D)
3. (A) (B) (C) (D)
4. (A) (B) (C) (D)
5. (A) (B) (C) (D)
6. GRID
7. GRID
8. (A) (B) (C) (D)
9. (A) (B) (C) (D)
10. (A) (B) (C) (D)

11. (A) (B) (C) (D)
12. (A) (B) (C) (D)
13. (A) (B) (C) (D)
14. (A) (B) (C) (D)
15. (A) (B) (C) (D)
16. (A) (B) (C) (D)

Use the answer spaces in the grids below if the question requires a grid-in response.

| Student-Produced Responses | ONLY ANSWERS ENTERED IN THE CIRCLES IN EACH GRID WILL BE SCORED. YOU WILL NOT RECEIVE CREDIT FOR ANYTHING WRITTEN IN THE BOXES ABOVE THE CIRCLES. |

6.

7.

PRACTICE SAT QUESTIONS

1. If $y = 4$ and $x = 3$, then $y^3 - 3x^2 + 3y - 2xy + x =$
   A. 25
   B. 26
   C. 27
   D. 28

2. $(x^2y - 4y^2 + 3xy^2) - (-x^2y + 5xy^2 - 3y^2)$

   Which of the following is equivalent to the expression above?
   A. $9x^2y^2$
   B. $-6xy^2 - 7y^2$
   C. $8xy^2 - 7y^2$
   D. $2x^2y - 2xy^2 - y^2$

3. $3(2x + 4) \times 4(2x + 4) = ax^2 + bx + c$

   In the equation above, a, b, and c are constants. If the equation is true for all values of x, what is the value of b?
   A. 48
   B. 96
   C. 192
   D. 200

4. $-(xy + x^3y - 7xy^3) - (x^3y + 4xy^3 - xy)$

   Which of the following choices is the same as the expression above?
   A. $-2xy + 3xy^3$
   B. $-2xy - 2x^3y + 3xy^3$
   C. $-2x^3y + 3xy^3$
   D. $2x^3y - 3xy^3$

5. $(4x^3 - 6) - (5x^3 - x + 6)$

   Which of the following is equivalent to the expression above?
   A. $-9x^3 - x - 12$
   B. $-x^3 + x - 12$
   C. $-9x^3 + x - 12$
   D. $9x^3 + x + 12$

6. When the expression below is written in the form $ax^2 + bx + c$, what is the value of c?

   $(4x^2 + 8x + 9) - 3(x^2 + 2x + 3)$

7. $(2x^2 + 4x + 7)$ subtracted from $(4x^2 + 5x - 2)$ is written in the form $ax^2 + bx + c$. What is the value of $(a + b) - c$?

8. $(1.5x - 2.2)^2 - (4.3x^2 - 2.8)$

   Which of the following is equivalent to the expression above?
   A. $3x^2 + 4.4x - 0.4$
   B. $2.05x^2 + 6.6x - 7.64$
   C. $-2.05x^2 - 6.6x + 7.64$
   D. $-2.05x^2 - 3.3x - 2.04$

9. $-(-2x^2 + 6) - (4x^2 + 9)$

   Which expression below is equivalent to the expression above?
   A. $-4x^2 - 15$
   B. $-2x^2 - 15$
   C. $6x^2 + 15$
   D. $-2x^2 + 15$

10. Which of the following is the sum of $(g - 1)$, $(g^2 + 3)$, and $(g - 2)$?
    A. $g^4 - 2g^3 - g^2 - 4g - 6$
    B. $g^2 + 2g$
    C. $2g^2 + g$
    D. $g^2 + 2g + 2$

11. The expression below is written in the form $ax^2 + c$. What is the value of $(a - c)$?

    $(3,450 + 90x^2) + 100(5x^2 - 200)$

    A. $-30,758$
    B. $-15,960$
    C. $15,960$
    D. $17,140$

12. $4(3x + 2)(5x + 2)$

    Which of the following is equivalent to the expression above?
    A. $60x^2 + 64x + 16$
    B. $22x^2 + 4$
    C. $60x$
    D. $60x^2 + 20$

13. $4x^8 + 16x^4y^4 + 4y^8$

    Which of the following is equivalent to the expression shown above?

    A. $2(8x^4 + 4y^4)^2$
    B. $4x^8 + y^4(16x^4 + 4y^4)$
    C. $(2x^2 + 4y^2)^4$
    D. $4(x^4 + 2y)(x^4 + 2y)$

14. If $a^2 + b^2 = x$ and $ab = z$, which of the following is equivalent to $2x + 4z$?

    A. $4(a^2 + b^2)$
    B. $2(a + b)^2$
    C. $(2a + 2b)^2$
    D. $(4a + 4b)^2$

15. $(2x^3 - 3y)^2$

    Which of the following is equivalent to the expression above?

    A. $2x^6 - 3y^2$
    B. $2x^5 - 3y^2$
    C. $2x^6 - 6x^3y - 6$
    D. $4x^6 - 12x^3y + 9y^2$

16. Which of the following is equivalent to the expression $(x - \frac{y^2}{3})$?

    A. $3y^2 + x$
    B. $\frac{1}{3}(3x - y^2)$
    C. $3x^2 - y^2$
    D. $(x - 3y)^2$

EXPLAINED ANSWERS

1. **EXPLANATION: Choice D is correct.**

    Substitute and solve.

    $y = 4$ and $x = 3$.

    $y^3 - 3x^2 + 3y - 2xy + x = 4^3 - 3(3)^2 + 3(4) - 2(3)(4) + 3 =$

    $64 - 27 + 12 - 24 + 3 = 28$.

2. **EXPLANATION: Choice D is correct.**

    First, distribute the negative sign: $x^2y - 4y^2 + 3xy^2 + x^2y - 5xy^2 + 3y^2$. Then, combine like terms: $(x^2y + x^2y)(-4y^2 + 3y^2)(3xy^2 - 5xy^2)$. If you've combined correctly, you should get $2x^2y - 2xy^2 - y^2$, or choice D.

3. **EXPLANATION: Choice C is correct.**

    Start by distributing the coefficients to get $(6x + 12) \times (8x + 16)$. Now use FOIL to turn the equation into the standard form, as requested: $6x$ times $8x = 48x^2$, 16 times $6x = 96x$, 12 times $8x = 96x$, 12 times $16 = 192$. $48x^2 + 192x + 192$ is the equation. What we're looking for is the coefficient of the x term, which in this case is choice C.

4. **EXPLANATION: Choice C is correct.**

    To do this question correctly, you need to combine like terms. Start with distributing the negative signs: $-xy - x^3y + 7xy^3 - x^3y - 4xy^3 + xy$. Now, combine like terms: $(-xy + xy) + (-x^3y - x^3y) + (7xy^3 - 4xy^3)$. You'll end up with choice C. The other answers stem from possible errors in combining or in distribution.

5. **EXPLANATION: Choice B is correct.**

    Like the question before, this one just needs like terms combined. Start with distributing the negative sign: $4x^3 - 6 - 5x^3 + x - 6$. Now, combine terms: the first term needs to be $-x^3$, and if you use the smart process of elimination, that's actually as far as you need to go to answer the question, since only one answer choice has that as a first term.

6. **EXPLANATION: The correct answer is 0.**

    Start by distributing, as usual, and you'll get $4x^2 + 8x + 9 - 3x^2 - 6x - 9$. Then, combine like terms: $x^2 + 2x$, but the 9 and -9 cancel each other out, which means that if this were in the standard form as requested in the question, it would be $x^2 + 2x + 0$. Therefore, $c = 0$.

7. **EXPLANATION: The correct answer is 12.**

    Start by subtracting the first expression from the second: careful, don't go backwards! $(4x^2 + 5x - 2) - (2x^2 + 4x + 7)$. Then, distribute the negative and combine like terms: $2x^2 + x - 9$. $a + b = 2 + 1 = 3$. $3 - c = 3 - (-9)$, or $3 + 9 = 12$.

8. **EXPLANATION: Choice C is correct.**

    Start by squaring the first parenthetical term, remembering to FOIL. $(1.5x - 2.2)(1.5x - 2.2) = 2.25x^2 - 3.3x - 3.3x + 4.84$. Next, simplify: $2.25x^2 - 6.6x + 4.84$. Now subtract the second parenthetical term from the original problem. $2.25x^2 - 6.6x + 4.84 - 4.3x^2 + 2.8$. The result is $-2.05x^2 - 6.6x + 7.64$, or choice C.

9. **EXPLANATION: Choice B is correct.**

    This problem requires distributing the negative signs. Start with distributing the negative to the first parentheses to get $2x^2 - 6 - (4x^2 + 9)$, then continue with the second parentheses: $2x^2 - 6 - 4x^2 - 9$. Now, combine terms as usual: $-2x^2 - 15$, or choice B.

10. **EXPLANATION: Choice B is correct.**

This problem is asking that you add the terms, so all you have to do is set it up as addition: $g - 1 + g^2 + 3 + g - 2$, and then combine and reorder the terms: $g^2 + 2g$ is all that is left because the addition of the numbers becomes 0 when completed.

11. **EXPLANATION: Choice D is correct.**

Start by combining like terms to get $3450 + 90x^2 + 500x^2 - 20,000$. Then, simplify: $590x^2 - 16,550$. The question asks us to subtract $a\,(590) - c\,(-16,550)$, so the answer is $(590) - (-16,550) = 590 + 16,550 = 17,140$. This is choice D.

12. **EXPLANATION: Choice A is correct.**

Start by using FOIL on the two parenthetical terms, which gives you $15x^2 + 16x + 4$. Then, multiply the entire expression by 4: $60x^2 + 64x + 16$.

13. **EXPLANATION: Choice B is correct.**

Start by factoring out a 4 from each term, so you now have $4(x^8 + 4x^4y^4 + y^8)$. Now consider the exponents: you can factor y^4 out of both the second and the third terms, and since the whole expression is addition, you can regroup the terms. Doing so gives you choice B. Alternatively, use your answer choices to work backwards.

14. **EXPLANATION: Choice B is correct.**

Plug in numbers for a and b: for example, make $a = 2$ and $b = 3$. In that case, $a^2 + b^2$ becomes $4 + 9 = 13$, so $x = 13$. $ab = 6$, so $z = 6$. Then, plug those numbers into the equation to solve: $2(13) + 4(6) = 50$. Plug the same numbers into the answer choices. For answer choice B, $2(2 + 3)^2 = 2(25) = 50$. You can use any numbers that fit the original two equations as a and b, and the problem will still work.

15. **EXPLANATION: Choice D is correct.**

Be sure you use FOIL to work through this problem fully—if you put one of the other choices, you likely made a mistake with rules of exponents. $(2x^3 - 3y)(2x^3 - 3y)$ makes the first term $4x^6$, the two middle terms are each $-6x^3y$, and the last term is $9y^2$.

16. **EXPLANATION: Choice B is correct.**

You cannot multiply an expression by a number. Therefore, you can eliminate choices A and C, and doing so will to steer you in that direction. That leaves you with B and D. Solve this expression by factoring out $\frac{1}{3}$, which gives choice B: the expression equals $\frac{1}{3}(3x - y^2)$.

CHAPTER 12

SOLVING AND EVALUATING FUNCTIONS

When solving and evaluating functions on the SAT, you might see the function as an equation. You might also see specific values of the function in a table or a graph. Begin with the seven worked out examples, which show you how to solve and evaluate functions. Then complete and correct the practice problems. Then review the 2 Solved SAT Problems and 11 Practice SAT Questions with answer explanations.

SOLVING AND EVALUATING FUNCTIONS

Example 1.

$f(x) = x^2 + 5$, $f(-2) =$

Substitute and solve.

$f(-2) = (-2)^2 + 5 = 4 + 5 = 9$.

Example 2.

$f(x) = 2^x + 5x$, $f(3) =$

Substitute and solve.

$f(3) = 2^3 + 5(3) = 8 + 15 = 23$.

Example 3.

$f(x) = x^3 - 4x$ and $g(x) = 2x + 5$, $f[g(2)] =$

Substitute.

$g(2) = 2(2) + 5 = 9$ $\Rightarrow$ $f(9) = 9^3 - 4 \times 9$ or $729 - 36 = 693$

Substitute −3 to $g(2)$ and solve.

$f[g(2)] = f(-3) = (-3)^3 - 4(-3)$

$-27 + 12 = -15$.

Example 4.

$f(x) = 6x^2 - 3x + 2$, $f(2x) = ?$

Solve: $f(2x) = 6(2x)^2 - 3(2x) + 2 = 6(4x^2) - 3(2x) + 2 = 24x^2 - 6x + 2$.

Example 5.

$f(x) = 2x - 6$, $f(a) = 8$, $a = ?$

Solve: $8 = 2a - 6$ $\Rightarrow$ $14 = 2a$ $\Rightarrow$ $7 = a$.

Example 6.

$f(x) = mx + 3$, $f(2) = 15$, $m = ?$

$15 = m \times 2 + 3$ $\Rightarrow$ $15 = 2m + 3$ $\Rightarrow$ $12 = 2m$ $\Rightarrow$ $6 = m$.

Example 7.

You may be able to solve a function without seeing it.

Consider $f(x) = 4$.

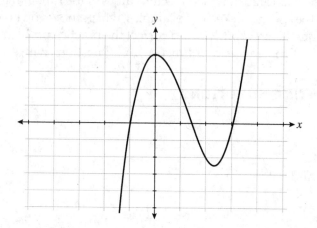

Inspect the graph and notice the function is equal to 4 when $x = 0$.

Practice Questions

1. $f(x) = 4x^3 - 5x^2 + 3x - 8$, $f(2) = ?$
2. $f(x) = 7x + 4$, $f(-4x^2) = ?$
3. $f(x) = -6x + 8$, $f(a) = -10$, $a = ?$
4. $f(x) = -2x^2 + 4x - 1$ and $g(x) = 5x - 7$, $f[g(2)] = ?$
5. Use the graph of $f(x)$ below to find $f(-2)$.

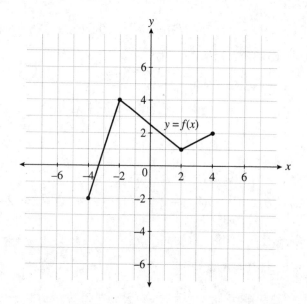

Practice Answers

1. $f(2) = 10$.
2. $f(-4x^2) = -28x^2 + 4$.
3. $a = 3$.
4. $f[g(2)] = -7$.
5. $f(-2) = 4$.

SOLVED SAT PROBLEMS

1. The table below shows specific values for functions $f(x)$ and $g(x)$. Use the table of values to compute $f[g(2)]$.

x	2	3	4	5
$f(x)$	3	8	15	24
$g(x)$	5	8	11	14

A. 3
B. 5
C. 8
D. 24

EXPLANATION: Choice D is correct.

First use the table to compute $g(2)$: $g(2) = 5$.

Next, use the table to compute $f[g(2)] = f(5)$: $f(5) = 24$.

2. The complete graph of the function f is shown in the graph below. What is the maximum value of $f(x)$?

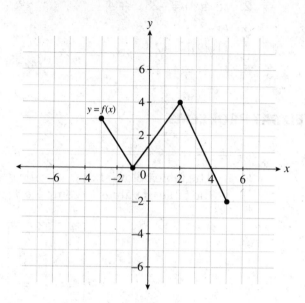

EXPLANATION: The correct answer is 4.

The question is asking for the maximum value of $f(x)$, which is $y = 4$.

The common error on this type of problem is to give an answer of 2, which is the answer to the question "For which value of x does $f(x)$ have a maximum?"

SOLVING AND EVALUATING FUNCTIONS
PRACTICE SAT QUESTIONS

ANSWER SHEET

Choose the correct answer.
If no choices are given, grid the answers in the section at the bottom of the page.

1. (A) (B) (C) (D)
2. (A) (B) (C) (D)
3. (A) (B) (C) (D)
4. GRID
5. (A) (B) (C) (D)
6. (A) (B) (C) (D)
7. (A) (B) (C) (D)
8. (A) (B) (C) (D)
9. (A) (B) (C) (D)
10. (A) (B) (C) (D)

11. (A) (B) (C) (D)

Use the answer spaces in the grids below if the question requires a grid-in response.

| Student-Produced Responses | ONLY ANSWERS ENTERED IN THE CIRCLES IN EACH GRID WILL BE SCORED. YOU WILL NOT RECEIVE CREDIT FOR ANYTHING WRITTEN IN THE BOXES ABOVE THE CIRCLES. |

4.

	(/)	(/)	(/)
	(.)	(.)	(.)
	(0)	(0)	(0)
(1)	(1)	(1)	(1)
(2)	(2)	(2)	(2)
(3)	(3)	(3)	(3)
(4)	(4)	(4)	(4)
(5)	(5)	(5)	(5)
(6)	(6)	(6)	(6)
(7)	(7)	(7)	(7)
(8)	(8)	(8)	(8)
(9)	(9)	(9)	(9)

PRACTICE SAT QUESTIONS

1. If $f(x) = \dfrac{x^2 - 4x + 2}{x + 2}$, what is $f(-4)$?

   A. −32
   B. −17
   C. 17
   D. 32

2. There are two functions, f and g.

   $g(4) = 8$ $g(6) = 12$ $f(8) = 9$ $f(6) = 4$

   What is $g[f(6)]$?

   A. 4
   B. 6
   C. 8
   D. 12

3. $f(x) = \dfrac{x^2 + 8x + 15}{x + 5}$ for $x \ne 5$, $f(-3) =$

   A. −9
   B. −3
   C. 0
   D. 3

4. $f(5) = 15$ and $g(x) = f(x + 2) - 5$. $g(3) =$

5.

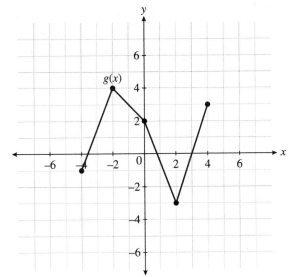

   The graph above shows all the plots of $g(x)$, a function on the Cartesian plane. For which value of x shown below is $g(x)$ at its maximum?

   A. −2
   B. 0
   C. 2
   D. 4

6.

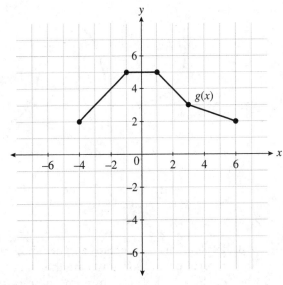

All the plots of the function $g(x)$ are shown on the graph above. For which values of x shown below is $g(x) \ge 3$?

   I. $-\dfrac{1}{2}$

   II. 3

   III. 6

   A. I only
   B. I and II only
   C. II and III only
   D. I, II, and III

7.

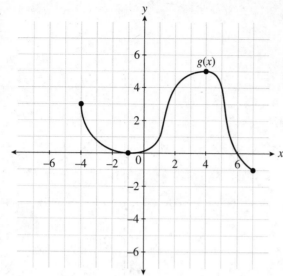

All the plots for the function $g(x)$ are shown on the graph above. Also shown are the values for the function $h(x)$. The minimum value of g is called m. What is $h(m)$?

x	0	1	2	3	4	5
$h(x)$	0	2	4	6	8	10

A. −2
B. 0
C. 2
D. 4

8.

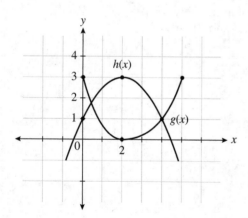

The graphs above show the plots of functions $g(x)$ and $h(x)$. For which of the following values of x does $g(x) - h(x) = -3$?

A. 0
B. 1
C. 2
D. 3

9. $g(x) = -4x - 3$. What is $g(-4x)$?

A. $16x + 3$
B. $16x - 3$
C. $-16x + -12$
D. $16x^2 + 12x$

10. $g(x) = ax^3 - 28$, where a is a positive constant.

If $g(-6) = -12$, what is the value of $g(6)$?

A. −44
B. 0
C. 12
D. 18

11.

x	$f(x)$
−4	12
3	2
6	0
10	−1

The function $f(x)$ is defined by a polynomial. Some values of x and $f(x)$ are shown in the table above. Which of the following must be a factor of $f(x)$?

A. $x + 4$
B. $x - 3$
C. $x - 6$
D. $x + 1$

EXPLAINED ANSWERS

1. **EXPLANATION: Choice B is correct.**

   Plug –4 in everywhere you see x: $(-4)^2 - 4(-4) + 2 = 34$ over $-4) + 2 = -2 = -17$. If you put another answer choice, you may have confused your signs. This is something the SAT will often use to trick you in functions questions, so look out for it!

2. **EXPLANATION: Choice C is correct.**

   This question is less complicated than it looks! $f(6) = 4$, according to what is given in the problem, so what you're actually looking for is $g(4)$, which the problem also gave us: $g(4) = 8$. The question gives extra information, which is common, but don't forget that this can happen. Not every item of information in a question must be used.

3. **EXPLANATION: Choice C is correct.**

   It is best to realize that the numerator can be factored.

   $$f(x) = \frac{x^2 + 8x + 15}{x + 5} \Rightarrow f(x) = \frac{(x + 5)(x + 3)}{x + 5}$$

   Simplify:

   $$f(x) = x + 3$$

   Substitute –3 for x.

   $$f(-3) = -3 + 3 = 0.$$

4. **EXPLANATION: The correct answer is 10.**

   So $g(x) = f(x + 2) - 5$

   Substitute 3 for x. $g(3) = f(3 + 2) - 5$

   Substitute 15 for $f(5)$. $f(5) - 5 = 15 - 5 = 10$.

   That means $g(3) = 10$.

5. **EXPLANATION: Choice A is correct.**

   A function is at its maximum when the y-value is highest. The y-value of $g(x)$ is highest when $x = -2$, according to the graph, so that's all you need to do to answer a question of this type.

6. **EXPLANATION: Choice B is correct.**

   Insert the values into the function to find the answer:

   I. $g\left(-\dfrac{1}{2}\right) = 5$ which is ≥ 3, so this is a potential answer. After checking this, you can eliminate choice C.

   II. $g(3) = 3$, which is ≥ 3. This one works as well. Now eliminate choice A.

   III. $g(6) = 2$, which is not ≥ 3. This one doesn't work, so the answer must be choice B, not choice D.

7. **EXPLANATION: Choice A is correct.**

   Inspection of the graph reveals that the minimum of $g(x)$ is –1. The table shows that the pattern we are following is to multiply x by 2 to find $h(x)$. $h(-1) = 2\,(-1) = -2$.

8. **EXPLANATION: Choice C is correct.**

Looking at the graph, it's clear that when $x = 2$, $g(x) = 0$, and $h(x) = 3$. $0 - (3) = -3$, so that's the answer you're looking for—the value of x needs to be 2. Choice A is incorrect, because it's the $g(x)$ value, and choice D is incorrect because it means you flipped the values for $g(x)$ and $h(x)$ in your evaluation. If you incorrectly assigned the $g(x)$ value as 1, you may have gotten choice B, which is also incorrect. Only choice C yields the correct value.

9. **EXPLANATION: Choice B is correct.**

In order to find $g(-4x)$, we need to first understand that we need to plug in $-4x$ whenever there is an x in the following equation: $g(x) = -4x - 3$. Therefore, when we plug it into the equation, we will get the following: $g(x) = -4(-4x) - 3$. This will result in $g(x) = 16x - 3$, which is choice B. The other answer choices can result if you incorrectly thought the question was asking you to solve for $-4g(x)$.

10. **EXPLANATION: Choice A is correct.**

Start by plugging in -6 for x to solve for a. $(-6)^3 a - 28 = -12$. So, $-216a - 28 = -12$. Simplify: $-216a = 16$. Divide to find that $a = -0.074074$, or $-\dfrac{2}{27}$. Now, use this information to solve the original question by plugging (6) in for x to find $g(6)$. $-\dfrac{2}{27}(6)^3 - 28 =$ your answer. To simplify, start with the exponent: $-\dfrac{2}{27}(216) = -16 - 28 = -44$, or choice A.

11. **EXPLANATION: Choice C is correct.**

Look at where $f(x) = 0$: this is a zero of the function. To get to an x of 6, one of the expressions when the functioned is factored must be $x - 6 = 0$, which indicates choice C.

CHAPTER 13

SOLVING QUADRATIC EQUATIONS

Quadratic equations on the SAT can sometimes be solved by taking the square root of both sides. Usually, you will write the equation in quadratic form ($ax^2 + bx + c = 0$). Then you may be able to factor the quadratic equation and set each factor equal to zero. You might need to use the **quadratic formula** ($ax^2 + bx + c = 0$) to find the solution. You will also see systems that will require substitution to create a single quadratic equation and then solve using one of the techniques described above. Begin with the mathematics review and then complete and correct the practice problems. There are 2 Solved SAT Problems and 11 Practice SAT Questions with answer explanations.

SOLVING QUADRATIC EQUATIONS

Quadratic equations take many different forms and may be hard to spot.

Example 1.

Solve the quadratic equation $(x - 4)^2 = 25$.

Take square root of both sides: $x - 4 = \pm\sqrt{25} = \pm 5$

Set $x - 4$ equal to 5 and solve: $x - 4 = 5 \Rightarrow x = 9$

Set $x - 4$ equal to -5 and solve: $x - 4 = -5 \Rightarrow x = -1$.

Example 2.

$x^2 + 12 = 7x \Rightarrow$ Write in quadratic form: $x^2 - 7x + 12 = 0$

Factor: $(x - 3)(x - 4) = 0$

Solve $x - 3 = 0$ and $x - 4 = 0 \Rightarrow x = 3$ and $x = 4$.

Example 3.

$2x^2 - 13x - 20 = 25 \Rightarrow$ Write in quadratic form: $2x^2 - 13x - 45 = 0$

Factor: $(2x + 5)(x - 9) = 0$

Solve $2x + 5 = 0$ and $x - 9 = 0 \Rightarrow x = -\dfrac{5}{2}$ and $x = 9$

A **quadratic equation** of the form $ax^2 + bx + c = 0$ can always be solved using the quadratic formula $x = \dfrac{-b \pm \sqrt{b^2 - 4ac}}{2a}$. This formula can be used to solve any quadratic equation in the form $ax^2 + bx + c = 0$, and it is the best technique when a quadratic equation cannot be factored.

Example 4.

Solve the quadratic equation $x^2 + 6x + 4 = 0$.

This equation cannot be factored, so you must use the quadratic formula.

Substitute $a = 1$, $b = 6$, and $c = 4$:

$$x = \frac{-b \pm \sqrt{b^2 - 4ac}}{2a} = \frac{-(6) \pm \sqrt{(6)^2 - 4(1)(4)}}{2(1)}$$

Simplify:

$$x = \frac{-6 \pm \sqrt{36 - 16}}{2} = \frac{-6 \pm \sqrt{20}}{2} = \frac{-6 \pm \sqrt{(4)(5)}}{2}$$

$$= \frac{-6 \pm 2\sqrt{5}}{2} = -3 \pm \sqrt{5}.$$

Example 5.

Solve the quadratic equation $3x^2 + 4x - 8 = 0$.

This problem cannot be factored, so you must use the quadratic formula.

Substitute $a = 3$, $b = 4$, and $c = -8$:

$$x = \frac{-b \pm \sqrt{b^2 - 4ac}}{2a} = \frac{-(4) \pm \sqrt{(4)^2 - 4(3)(-8)}}{2(3)}$$

Simplify:

$$x = \frac{-4 \pm \sqrt{16 + 96}}{6} = \frac{-4 \pm \sqrt{112}}{6} = \frac{-4 \pm \sqrt{(16)(7)}}{6} = \frac{-4 \pm 4\sqrt{7}}{6} = \frac{-2 \pm 2\sqrt{7}}{3}.$$

Example 6.

Solve the quadratic equation $2x^2 - 16x = 4$.

Write the equation in Quadratic Form. Subtract 4 from both sides of the equations: $2x^2 - 16x - 4 = 0$

Factor out a 2: $2(x^2 - 8x - 2) = 0$

Divide both sides of equation by 2: $x^2 - 8x - 2 = 0$

This equation cannot be factored, so you must use the quadratic formula.

Substitute $a = 1$, $b = -8$, and $c = -2$:

$$x = \frac{-b \pm \sqrt{b^2 - 4ac}}{2a} = \frac{-(-8) \pm \sqrt{(-8)^2 - 4(1)(-2)}}{2(1)}$$

Simplify: $x = \dfrac{8 \pm \sqrt{64 + 8}}{2} = \dfrac{8 \pm \sqrt{72}}{2} = \dfrac{8 \pm \sqrt{(36)(2)}}{2} = \dfrac{8 \pm 6\sqrt{2}}{2} = 4 \pm 3\sqrt{2}.$

Example 7.

Solve the system: $y = 4x^2 - 7x + 8$

$$y = 2x^2 + 3x - 4$$

Set the two equations equal: $4x^2 - 7x + 8 = 2x^2 + 3x - 4$

Subtract $2x^2$ from both sides: $2x^2 - 7x + 8 = 3x - 4$

Subtract $3x$ from both sides: $2x^2 - 10x + 8 = -4$

Add 4 to both sides: $2x^2 - 10x + 12 = 0$

Factor out a 2: $2(x^2 - 5x + 6) = 0$

Divide both sides by 2: $x^2 - 5x + 6 = 0$

Factor: $(x - 2)(x - 3) = 0$

Set each factor equal to 0 and solve: $(x - 2) = 0 \implies x = 2$

$$(x - 3) = 0 \implies x = 3.$$

Practice Problems

1. Solve the quadratic equation $(x + 5)^2 - 4 = 12$.
2. Solve the quadratic equation $x^2 - 7x = 30$.
3. Solve the quadratic equation $3x^2 + 27x + 60 = 0$.
4. Solve the quadratic equation $3x^2 + 12x + 3 = 0$.
5. Solve the system: $y = 5x^2 - 4x - 5$

$$y = 3x^2 - 5x + 10$$

Practice Answers

1. $x = -1$ and $x = -9$.
2. $x = 10$ and $x = -3$.
3. $x = -4$ and $x = -5$.
4. $x = -2 \pm \sqrt{3}$.
5. $x = \dfrac{5}{2}$ and $x = -3$.

SOLVED SAT PROBLEMS

---

1. If (x, y) is a solution of the system of equations below and $x < 0$, what is the value of x?

$$y = x^2 + 2x - 4$$
$$4y - 3 = 3(x^2 - 1)$$

A. $x = -4 - 4\sqrt{2}$
B. $x = -4 + 4\sqrt{2}$
C. $x = -2 - 2\sqrt{2}$
D. $x = -2 + 2\sqrt{2}$

EXPLANATION: Choice A is correct.

Substitute $x^2 + 2x - 4$ for y in the second equation:

$4(x^2 + 2x - 4) - 3 = 3(x^2 - 1)$

Distribute the 4 on left side of equation: $4x^2 + 8x - 16 - 3 = 3(x^2 - 1)$

Simplify left side of equation: $4x^2 + 8x - 19 = 3(x^2 - 1)$

Distribute the 3 on right side of equation: $4x^2 + 8x - 19 = 3x^2 - 3$

Subtract $3x^2$ from both sides of equation: $x^2 + 8x - 19 = -3$

Add 3 to both sides of equation: $x^2 + 8x - 16 = 0$

Substitute $a = 1$, $b = 8$, and $c = -16$:

$$x = \frac{-b \pm \sqrt{b^2 - 4ac}}{2a} = \frac{-(8) \pm \sqrt{(8)^2 - 4(1)(-16)}}{2(1)}$$

Simplify:

$$x = \frac{-8 \pm \sqrt{64 + 64}}{2} = \frac{-8 \pm \sqrt{128}}{2} = \frac{-8 \pm \sqrt{(64)(2)}}{2} = \frac{-8 \pm 8\sqrt{2}}{2}$$

$$= -4 \pm 4\sqrt{2}$$

Since $x < 0$, $x = -4 - 4\sqrt{2}$. Choice B is not correct because x would be greater than 0.

2. In the equation below, which of the following is a possible value of x?

$$\frac{3}{x-5} = \frac{x-5}{12}$$

A. 1
B. 11
C. 13
D. 23

EXPLANATION: Choice B is correct.

Cross multiply: $(x-5)^2 = 36$

Take the square root of both sides of the equation: $x - 5 = \pm\sqrt{36} = \pm 6$

Add 5 to both sides of the equation and simplify: $x = 5 \pm 6$

Therefore, either $x = -1$ or $x = 11$.

Based on answer choices, the correct answer is choice B.

Alternatively, you can test out the answer choices in the question.

SOLVING QUADRATIC EQUATIONS
PRACTICE SAT QUESTIONS

ANSWER SHEET

Choose the correct answer.
If no choices are given, grid the answers in the section at the bottom of the page.

1. Ⓐ Ⓑ Ⓒ Ⓓ
2. **GRID**
3. **GRID**
4. Ⓐ Ⓑ Ⓒ Ⓓ
5. Ⓐ Ⓑ Ⓒ Ⓓ
6. Ⓐ Ⓑ Ⓒ Ⓓ
7. Ⓐ Ⓑ Ⓒ Ⓓ
8. **GRID**
9. **GRID**
10. Ⓐ Ⓑ Ⓒ Ⓓ

11. Ⓐ Ⓑ Ⓒ Ⓓ

Use the answer spaces in the grids below if the question requires a grid-in response.

Student-Produced Responses　　ONLY ANSWERS ENTERED IN THE CIRCLES IN EACH GRID WILL BE SCORED. YOU WILL NOT RECEIVE CREDIT FOR ANYTHING WRITTEN IN THE BOXES ABOVE THE CIRCLES.

PRACTICE SAT QUESTIONS

1. How many solutions are there to the system of equations shown below?

$$y = 2x^2 - 8x + 6$$
$$y = 4x + 12$$

A. There are no solutions.
B. There is exactly one solution.
C. There are exactly two solutions.
D. There are exactly three solutions.

2. If the ordered pair (x, y) satisfies the system of equations below, what the value of y?

$$y = x^2 + 6$$
$$y = -x^2 + 6$$

3. What is the product of the possible solutions to the system of equations shown below?

$$y = x^2 - 4x$$
$$2y - 6x = -24$$

4. If $x^2 + 2xy + y^2 = 16$ and $x^2 - y^2 = 4$, which of the following are possible values of $x - y$?

   I. -1
   II. 0
   III. 1

A. II only
B. III only
C. I and III only
D. I, II, and III

5. What is the product of the solutions to $(x - 2)(x + 4) = 0$?

A. -8
B. -2
C. 2
D. 8

6. What are the solutions to the quadratic equation below?

$$x^2 - 2x - 24 = 0$$

A. $x = -6$ and $x = -4$
B. $x = -6$ and $x = 4$
C. $x = 6$ and $x = -4$
D. $x = 6$ and $x = 4$

7. What are the solutions to the quadratic equation seen below?

$$3x^2 - 18x + 24 = 0$$

A. $x = -4$ and $x = -2$
B. $x = -4$ and $x = 2$
C. $x = 4$ and $x = -2$
D. $x = 4$ and $x = 2$

8. In the equation $(x - 4)^2 - 9 = 0$, what is a possible value of x?

9. In the equation below, what is a possible value of x?

$$\frac{x - 4}{x} = \frac{2}{x - 4}$$

10. What are the solutions to $2x^2 + 12x - 6 = 0$?

A. $-3 \pm \sqrt{6}$
B. $-3 \pm 2\sqrt{6}$
C. $-3 \pm \sqrt{3}$
D. $-3 \pm 2\sqrt{3}$

11. The equation $4x^2 + 12x + c = 0$ is written in standard form and has two real solutions. Which of the following could be a value of c?

A. 8
B. 9
C. 10
D. 11

EXPLAINED ANSWERS

1. **EXPLANATION : Choice C is correct.**

Set the equations equal:	$2x^2 - 8x + 6 = 4x + 12$
Combine like terms:	$2x^2 - 12x - 6 = 0$
Divide by 2:	$x^2 - 6x - 3 = 0$
Test the discriminant with $a = 1, b = -6, c = -3$:	$(-6)^2 - 4\,(1)(-3)$
	$36 - (-12) = 48$

   Since the discriminant is positive, there are two solutions.

   Alternatively, the equations can be graphed if a calculator is permitted.

2. **EXPLANATION: The correct answer is 6.**

Set the two equations equal to each other:	$x^2 + 6 = -x^2 + 6$
Simplify by combining the x^2 values on the left and numbers on the right:	$2x^2 = 0; x = 0$
Now, substitute $x = 0$ into either parabola (the original equations):	$y = 0 + 6$
	$y = 6.$

3. **EXPLANATION: The correct answer is 12.**

   Start by putting the second equation into slope-intercept form. First, each number is divided by $2 \Rightarrow y - 3x = -12$.

Isolate the y-variable for the second equation:	$\Rightarrow y = 3x - 12$
Then, set the equations equal to each other:	$x^2 - 4x = 3x - 12$
Simplify and set the equation equal to zero:	$x^2 - 1x + 12 = 0$
Now, factor to get the following:	$(x - 4)\,(x - 3) = 0$
Then solve:	$x - 4 = 0$, so $x = 4$
	$x - 3 = 0$, so $x = 3$

   Therefore, the two values in question are 3 and 4: multiply them together to get 12, the answer.

4. **EXPLANATION: Choice C is correct.**

   Start by factoring the first equation: it factors to $(x + y)(x + y) = 16$. Simplifying both sides tells you that $(x + y) = 4$ or $(x + y) = -4$. Now, factor the second equation, which factors to $(x - y)(x + y)$. We know that $(x - y)(x + y) = 4$ or -4 and $(x + y) = 4$. Therefore, $(x - y)(4) = 4$ or $(x - y)(-4) = 4$. Thus, $(x - y)$ must be equal to 1 or -1.

5. **EXPLANATION: Choice A is correct.**

   Set each side of the equation equal to 0 to get the two possible solutions: $(x - 2) = 0$, so x equals 2. $(x + 4) = 0$, so x equals -4. Now, multiply the two together: $(2)(-4) = -8$.

6. **EXPLANATION: Choice C is correct.**

   This equation is easily factored, although not all equations on the SAT will be, so be ready to use the quadratic formula as needed. In this case, the equation factors to $(x - 6)(x + 4)$. Set both of these factors equal to 0 to get $x = 6$ and $x = -4$, or choice C.

7. **EXPLANATION: Choice D is correct.**

   Although simplifying by dividing before you factor is not necessary, it can be helpful. Divide each term by 3 to get $x^2 - 6x + 8 = 0$. Now, factor the equation: $(x - 4)(x - 2)$. Possible values for x are represented by choice D.

8. **EXPLANATION: The correct answer is 7, 1.**

   First, use FOIL to square the first part of the equation: $(x-4)(x-4) = x^2 - 8x + 16 - 9 = 0$. Combine like terms to get $x^2 - 8x + 7 = 0$. Now, factor: $(x-7)(x-1)$: possible values for x are therefore 7 and 1, and you can grid in either as your response to get credit.

9. **EXPLANATION: The correct answer is 8, 2.**

First, cross multiply:	$2x = (x-4)(x-4)$
Now, use FOIL:	$2x = x^2 - 8x + 16$
Subtract $2x$ from both sides:	$x^2 - 10x + 16$
Then factor:	$(x-8)(x-2)$
Then you can set them to 0:	$(x-8) = 0$

   So $x = 8$; $(x-2) = 0$, so $x = 2$.

   Thus, the possible solutions are 8 and 2, and you can grid either of these in as your answer to get credit.

10. **EXPLANATION: Choice D is correct.**

    Occasionally, the SAT will use an equation for which you must use the quadratic formula, because it cannot be factored. Start by dividing all the terms by 2: $x^2 + 6x - 3 = 0$. So, $a = 1$, $b = 6$, $c = -3$. Using the equation $x = \dfrac{-b\sqrt{b^2 - 4ac}}{2a}$, the equation becomes $\dfrac{-6 \pm \sqrt{6^2 - 4(1)(-3)}}{2}$. This reduces to $\dfrac{-6 \pm \sqrt{48}}{2}$. Now, simplify the root: $\sqrt{3} \times \sqrt{16} = 4\sqrt{3}$, so now the equation is $\dfrac{-6 \pm 4\sqrt{3}}{2}$. Divide a final time to get choice D.

11. **EXPLANATION: Choice A is correct.**

    In order to have two real solutions, the equation must cross the x-axis in two places. That means that the discriminant, $b^2 - 4ac$, must be greater than 0. Test out your answer choices. Only choice A gives a discriminant larger than 0 (choice B gives a discriminant equal to 0, which means only one real solution; the others have a negative discriminant, which means two imaginary solutions). To take this a step further, instead of testing solutions, you could set up an inequality: $b^2 - 4ac > 0$ or $144 - 16c > 0$, and find that $c < 9$, leaving only A as an option.

CHAPTER 14

GRAPHS OF QUADRATIC FUNCTIONS

The graph of a quadratic function is a **parabola**. A parabola has two key features: (1) the *x*-intercepts, where the function (*y*) value is zero, and (2) the **vertex**, which is either the largest or smallest value of the function. The SAT will give you different forms of a quadratic function. You will need to determine if the given form is useful or if it needs to be manipulated to answer the question. Other SAT questions will indicate that a point is on the graph of a parabola and ask you to find *a*, *b*, or *c* in the standard formula for that parabola. Other questions ask you to find the intersection of a parabola and linear functions. Begin with the mathematics review and then complete and correct the practice problems. There are 2 Solved SAT Problems and 10 Practice SAT Questions with answer explanations.

GRAPHS OF QUADRATIC FUNCTIONS

Example 1.

Identify the *x*-intercepts for the graph of the quadratic function $f(x) = (x - 6)(x + 3)$.

Set each factor equal to 0 and solve: $(x - 6) = 0 \rightarrow x = 6$

$(x + 3) = 0 \rightarrow x = -3$.

Example 2.

Identify the *x*-intercepts for the graph of the quadratic function $f(x) = x^2 - 8x + 12$.

Factor: $f(x) = (x - 6)(x - 2)$

Set each factor equal to 0 and solve: $(x - 6) = 0 \rightarrow x = 6$

$(x - 2) = 0 \rightarrow x = 2$.

Example 3.

Find the *x*-value of the vertex of the parabola with equation $f(x) = (x + 5)(x - 3)$.

The *x*-value of the vertex is always the midpoint of the *x*-intercepts.

Set each factor equal to 0 and solve: $(x + 5) = 0 \rightarrow x = -5$

$(x - 3) = 0 \rightarrow x = 3$

To find the midpoint add the *x* values and divide by 2.

The midpoint of $x = -5$ and $x = 3 \Rightarrow -5 + (3) = -2 \Rightarrow \dfrac{-2}{2} = -1$ is $x = -1$.

Therefore, the *x*-value for the vertex is $x = -1$.

Example 4.

Find the *x*-value of the vertex of the parabola with equation $f(x) = x^2 - 6x - 27$.

The *x*-value of the vertex is always the midpoint of the *x*-intercepts.

Factor: $f(x) = (x - 9)(x + 3)$

Set each factor equal to 0 and solve: $(x - 9) = 0 \;\;\rightarrow\;\; x = 9$

$\qquad\qquad\qquad\qquad\qquad\qquad\quad (x + 3) = 0 \;\;\rightarrow\;\; x = -3$

To find the midpoint add the x values and divide by 2.

The midpoint of $x = 9$ and $x = -3 \;\;\Rightarrow\;\; 9 + (-3) = 6 \;\;\Rightarrow\;\; \dfrac{6}{2} = 3$ is $x = 3$.

Therefore, the x-value for the vertex is $x = 3$.

Example 5.

Find the coordinates of the vertex for the graphs of the quadratic functions below and determine if it is a maximum or a minimum.

A. $f(x) = -2(x - 4)^2 + 5$

B. $f(x) = (x + 2)^2 - 6$

C. $f(x) = -(x - 3)^2 - 1$

D. $f(x) = 3(x + 7)^2 + 5$.

This is called the *vertex form* of a quadratic function because the vertex can be quickly identified.

A. The vertex is (4, 5). Since the function has a negative leading coefficient of –2, it is opening down, so the vertex of (4, 5) is a maximum.

B. The vertex is (–2, –6). Since the function has a positive leading coefficient of 1, it is opening up, so the vertex of (–2, –6) is a minimum.

C. The vertex is (3, –1). Since the function has a negative leading coefficient of –1, it is opening down, so the vertex of (3, –1) is a maximum.

D. The vertex is (–7, 5). Since the function has a positive leading coefficient of 3, it is opening up, so the vertex of (4, 5) is a maximum.

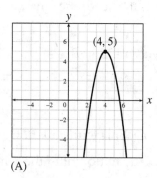

(A)

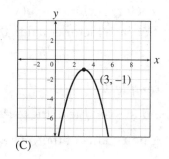

(C)

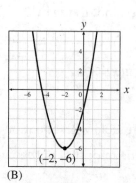

(B)

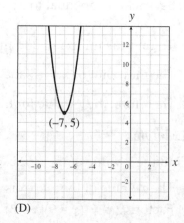

(D)

Example 6.

Convert the quadratic function $f(x) = x^2 + 6x - 3$ into vertex form and give the coordinate of the vertex.

Leave space between constant and variables: $\qquad f(x) = x^2 + 6x \qquad -3$

Square half of the x-coefficient: $\qquad \dfrac{6}{2} = 3 \quad \rightarrow \quad (3)^2 = 9$

Add **9** and subtract **9** to balance out equation: $\qquad f(x) = \underbrace{x^2 + 6x + 9} \quad -3 - 9$

Factor the perfect square in bracket and simplify: $\qquad f(x) = (x + 3)^2 \qquad -12$

The vertex form is $f(x) = (x + 3)^2 - 12$ and the vertex is $(-3, -12)$.

Example 7.

Convert the quadratic function $f(x) = x^2 - 8x + 22$ into vertex form and give the coordinate of the vertex.

Leave space between constant and variables: $\qquad f(x) = x^2 - 8x \qquad + 22$

Square half of the x-coefficient: $\qquad \dfrac{-8}{2} = -4 \quad \rightarrow \quad (-4)^2 = 16$

Add **16** and subtract **16** to balance out equation: $\quad f(x) = \underbrace{x^2 - 8x + 16} \quad + 22 - 16$

Factor the perfect square in bracket and simplify: $\qquad f(x) = (x - 4)^2 \qquad + 6$

The vertex form is $f(x) = (x - 4)^2 + 6$ and the vertex is $(4, 6)$.

Example 8.

The point $(2, 12)$ lies on the graph of $f(x) = 5x^2 - 3x + c$. What is the value of c?

$$12 = 5(2)^2 - 3(2) + c \quad \rightarrow \quad 12 = 20 - 6 + c$$

$$12 = 14 + c \quad \rightarrow \quad -2 = c.$$

Example 9.

The point $(-3, 5)$ lies on the graph of $f(x) = ax^2 - 6x - 14$. What is the value of a?

$$5 = a(-3)^2 - 6(-3) - 14 \quad \rightarrow \quad -5 = 9a + 18 - 14$$

$$-5 = 9a + 4 \quad \rightarrow \quad -9 = 9a \quad \rightarrow \quad -1 = a.$$

Practice Questions

1. Identify the x-intercepts for the graph of the quadratic function $f(x) = x^2 - 2x - 35$.
2. Find the x-value of the vertex of the parabola with equation $f(x) = x^2 - 8x - 20$.
3. Find the coordinates of the vertex for the graph of $f(x) = -4(x + 7)^2 + 3$ and determine if it is a maximum or a minimum.
4. Convert the quadratic function $f(x) = x^2 - 10x + 8$ into vertex form and give the coordinate of the vertex.
5. The point $(-2, 10)$ lies on the graph of $f(x) = 3x^2 + bx + 8$. What is the value of b?

Practice Answers

1. $x = 7$ and $x = -5$.
2. $x = 4$.
3. The vertex of $(-7, 3)$ is a maximum.
4. $f(x) = (x - 5)^2 - 17$ and the vertex is $(5, -17)$.
5. $b = 5$.

SOLVED SAT PROBLEMS

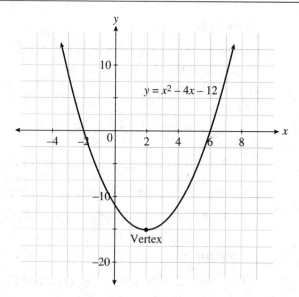

1. Which of the following is an equivalent form of the equation of the graph shown above from which the x-intercepts of the parabola can be identified as constants in the equation?

   A. $y = (x + 2)^2 - 16$
   B. $y = (x - 2)^2 - 16$
   C. $y = (x - 2)(x + 6)$
   D. $y = (x + 2)(x - 6)$

   EXPLANATION: Choice D is correct.

   Factoring the original equation gives $y = x^2 - 4x - 12 = (x + 2)(x - 6)$.

   From this equation it can be seen that the x-intercepts of the parabola are $x = -2$ and $x = 6$.

2. The equation below represents a parabola in the xy-plane. Which of the following is an equivalent form of the equation that displays the vertex of the parabola as constants?

$$f(x) = x^2 + 2x - 8$$

A. $y = (x - 1)^2 - 8$
B. $y = (x + 1)^2 - 9$
C. $y = (x - 2)(x + 4)$
D. $y = (x + 2)(x - 4)$

EXPLANATION: Choice B is correct.

Leave space between constant and variables: $\qquad f(x) = x^2 + 2x \qquad - 8$

Square half of the x-coefficient: $\qquad \dfrac{2}{2} = 1 \quad \rightarrow \quad (1)^2 = 1$

Add **16** and subtract **16** to balance out equation:
$$f(x) = \underbrace{x^2 + 2x + 1} - 8 - 1$$

Factor the perfect square in bracket and simplify: $\qquad f(x) = (x + 1)^2 \quad - 9$

From this form it can be seen that the vertex is $(-1, 9)$.

GRAPHS OF QUADRATIC FUNCTIONS
PRACTICE SAT QUESTIONS

▨ **ANSWER SHEET**

Choose the correct answer.
If no choices are given, grid the answers in the section at the bottom of the page.

1. Ⓐ Ⓑ Ⓒ Ⓓ
2. Ⓐ Ⓑ Ⓒ Ⓓ
3. Ⓐ Ⓑ Ⓒ Ⓓ
4. Ⓐ Ⓑ Ⓒ Ⓓ
5. Ⓐ Ⓑ Ⓒ Ⓓ
6. Ⓐ Ⓑ Ⓒ Ⓓ
7. Ⓐ Ⓑ Ⓒ Ⓓ
8. Ⓐ Ⓑ Ⓒ Ⓓ
9. Ⓐ Ⓑ Ⓒ Ⓓ
10. GRID

Use the answer spaces in the grid below if the question requires a grid-in response.

Student-Produced Responses	ONLY ANSWERS ENTERED IN THE CIRCLES IN EACH GRID WILL BE SCORED. YOU WILL NOT RECEIVE CREDIT FOR ANYTHING WRITTEN IN THE BOXES ABOVE THE CIRCLES.

10.

PRACTICE SAT QUESTIONS

1.

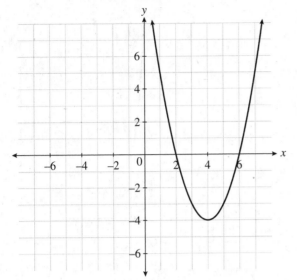

The graph above shows the equation $x^2 - 8x + 12$. Which of the following is an identical equation that represents the graph above from which the coordinates of the vertex can be determined?

A. $y = x(x - 8) + 12$
B. $y = (x - 4)^2 - 4$
C. $y = (x - 6)(x + 2)$
D. $y = (x - 6)(x - 2)$

2.

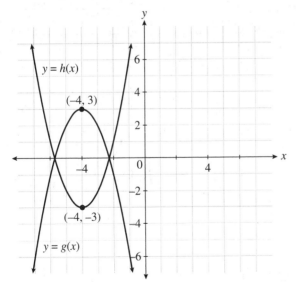

The graph shows functions $g(x)$ and $h(x)$, which represent the paths of two balls in the air that Susan is using in an experiment. Susan needs to determine for which value of x, do $g(x)$ and $h(x)$ represent a number and its negative value.

A. -4
B. 0
C. 3
D. 4

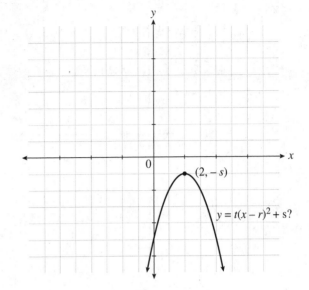

3. Which of the following statements about the characteristics of the parabola is factual, given that it represents the equation $y = t(x - r)^2 + s$? The parabola is shown above.

A. The vertex is (r, s) and the graph opens upward.
B. The vertex is (r, s) and the graph opens downward.
C. The vertex is $(-r, s)$ and the graph opens upward.
D. The vertex is $(-r, s)$ and the graph opens downward.

4.

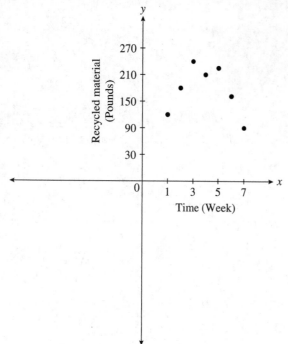

The scatterplot shows the amount of material recycled by Joe's neighborhood, in hundreds of pounds, over a period of several weeks. Which equation is the best model for the data in Joe's scatter plot?

A. $7.67x^2 + 53.03x + 87.8$
B. $-7.67x^2 - 53.03x - 87.8$
C. $-7.67x^2 + 53.03x + 87.8$
D. $7.67x^2 + 53.03x + 87.8$

5. $y = x^2 + 4x - 12$
$y = 2x + 12 - 0$

How many solutions are there to the system of equations above?

A. There are no solutions.
B. There are two solutions.
C. There are four solutions.
D. There are five solutions.

6. The zeros of a polynomial function are –2 and 1. In the xy-plane, which of the following graphs represents this function most accurately?

A.

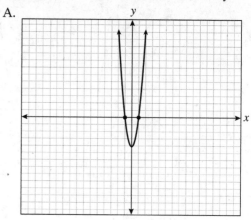

B.

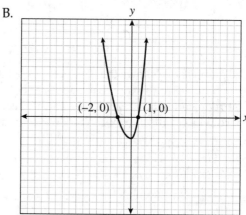

C.

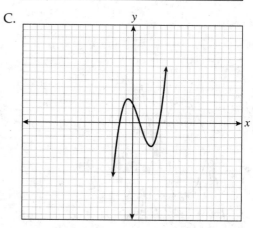

D.

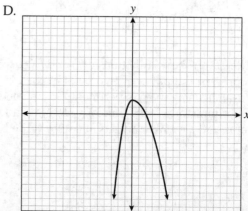

7.

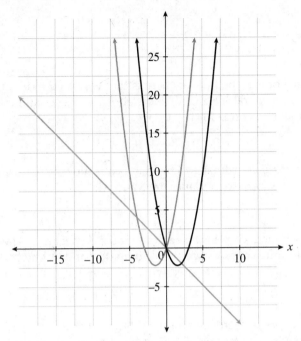

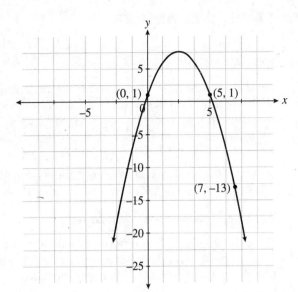

The xy-plane above shows three distinct equations, represented by this system: $\begin{cases} y = x^2 + 3x \\ y = -x \\ y = x^2 - 3x \end{cases}$.

For this xy-plane, with two parabolas and one line, how many solutions does the system graphed here contain?

A. Exactly three
B. Exactly two
C. Exactly one
D. No solutions

8. The graph above is a parabola. Which of the following defines the function?

A. $(x + 2.25)^2 + 7.5$
B. $-(x - 2.25)^2 + 7.5$
C. $(x + 2.25)^2 - 7.5$
D. $-(x - 2.25)^2 - 7.5$

9. The graph of function c crosses the x-axis at two points, $(d, 0)$ and $(e, 0)$. If we know that d is negative and e is positive, which of the following could define function c, where a and b are positive constants?

A. $c(x) = (x + a)(x + b)$
B. $c(x) = x(x - a)(x - b)$
C. $c(x) = (x - a)(x - b)$
D. $c(x) = (x + a)(x - b)$

10. $x^2 + 4x + 6$

The equation above represents a parabola. The vertex of the parabola is represented by (c, d). What is the value of d?

EXPLAINED ANSWERS

1. **EXPLANATION: Choice B is correct.**

   Several of these answer choices are equivalent forms of the line; the key here is that you need to identify which of those has the vertex in it. Note the vertex of the parabola: (4, −4). Now, look for the equation that identifies those coordinates within it: it's choice B that's in vertex form.

2. **EXPLANATION: Choice A is correct.**

   For $g(x)$ and $h(x)$ to equal a number and its negation, the y-coordinates of the two points must be opposites. Looking at the graph, it's clear that this happens at −4, where the x values are 3 and −3.

3. **EXPLANATION: Choice B is correct.**

   Since the parabola shown opens downward, the coefficient of x^2 in the equation must be negative. Therefore, t is positive, and the parabola the answer calls for must open upward, so eliminate B and D. The vertex of the parabola will be (r, s) because the minimum value of y, s, is reached when x is equal to r.

4. **EXPLANATION: Choice C is correct.**

   Joe's scatter plot is a rough parabola that opens down, so the coefficient for the x^2 term needs to be negative: eliminate A and B. The y-intercept of the parabola needs to be around 90, so the equation that gives the best approximation is choice C.

5. **EXPLANATION: Choice B is correct.**

   The first equation is a parabola. The second is a line. A parabola and a line can intersect at no more than two points. Solving the two equations (or just graphing them on your calculator) will demonstrate that they intersect at two points.

6. **EXPLANATION: Choice B is correct.**

   The range of p gives all the possible y values, so the function can only have points with y-values less than or equal to 3. If the zeros are −2 and 1, that is where it needs to cross the x-axis. Choice B meets these qualifications.

7. **EXPLANATION: Choice C is correct.**

   The solution to any system of equations must be a solution to all of the equations in the system. In this case, the parabola and both lines intersect all together at only one point, so there is only one solution.

8. **EXPLANATION: Choice B is correct.**

   The equation of a parabola in vertex form is $f(x) = a(x − h)^2 + k$, and the point (h, k) should represent the vertex. In this case, the vertex is at approximately (2.5, 7.25), so the equation is $a(x − 2.5)^2 + 7.25$. To find a, substitute any of the points given into the equation. For example, $1 = a(0 − 2.5)^2 + 7.25$. Simplify to get $1 = 6.25a + 7.25$. $a = −1$. Therefore, the equation can be written $−(x − 2.25)^2 + 7.25$, choice B.

9. **EXPLANATION: Choice D is correct.**

   Since the equation crosses the x-axis at those two points, those are the zeros of the equation. If we plug in $d = −2$ and $e = 3$, then $c(x) = (x − (−2))(x − 3)$, or $(x + 2)(x − 3)$. Look for the equation that uses these two signs, choice D.

10. **EXPLANATION: The correct answer is 2.**

    Find the vertex form of the parabola using the following steps:

    $f(x) = x^2 + 4x + 6$

    $f(x) = x^2 + 4x + 2^2 − 2^2 + 6$ (Complete the square.)

    $f(x) = (x + 2)^2 + 2$ (Simplify.)

    The vertex form has the x-coordinate, in our case c, in the parentheses in the form $(x − c)$ and the y-coordinate outside of the parentheses, 2.

CHAPTER 15

POLYNOMIALS, FACTORS, AND ROOTS

On the SAT you will associate the x – intercepts (roots or zeros) of a polynomial with the factors of the polynomial. Specifically, if $x = a$ is a **root** of a polynomial, then $x - a$ is a **factor** of the polynomial. In addition, you will need to connect the equation of a polynomial written as a product of the factors to a possible graph of the polynomial. Begin with the mathematics review and then complete and correct the practice problems. There are 2 Solved SAT Problems and 5 Practice SAT Questions with answer explanations.

POLYNOMIALS, FACTORS, AND ROOTS

Example 1.

Given the roots of a polynomial are –4, –2, and 7, write a possible equation of the polynomial as a product of its factors.

Since $x = -4$ is a root, then $x + 4$ is a factor.

Since $x = -2$ is a root, then $x + 2$ is a factor.

Since $x = 7$ is a root, then $x - 7$ is a factor.

A possible equation of the polynomial is $y = (x + 4)(x + 2)(x - 7)$.

Example 2.

Write a possible equation of the graph below as a product of its factors.

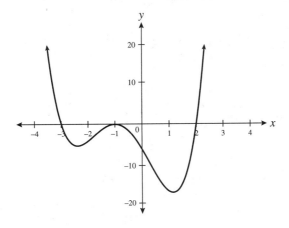

When you look at the graph it seems the roots are –3, –1, and 2, so the factors are $(x + 3)$, $(x + 1)$, and $(x - 2)$.

However, notice that at $x = -1$ the graph does not go through the x-axis but appears to bounce off the x-axis. When a graph just touches the x-axis there is a double root. That means this graph has a double root at $x = -1$. The roots for this graph are –3, –1, –1 and 2. So the factors are $(x + 3)$, $(x + 1)$, $(x + 1)$, and $(x - 2)$, which can be written as $(x + 3)(x + 1)^2 (x - 2)$.

Example 3.

Find the x-intercepts of the polynomial $y = x^3 + 2x^2 - 9x - 18$.

We want to factor this equation, but since it is not a quadratic equation, there is a little more to the factoring process.

First separate the terms into binomials: $y = \underbrace{x^3 + 2x^2}\ \underbrace{- 9x - 18}$

Factor out what you can from each binomial: $y = x^2 \underbrace{(x + 2)} - 9 \underbrace{(x + 2)}$

Since $(x + 2)$ is common to both binomials, factor out $(x + 2)$: $y = (x + 2)(x^2 - 9)$.

$(x^2 - 9)$ can be factored further as $(x - 3)(x + 3)$: $y = (x + 2)(x - 3)(x + 3)$.

The x-intercepts of the polynomial are $x = -2, x = 3$, and $x = -3$.

Example 4.

Find the zeros of the polynomial $y = x^4 - 13x^2 + 36$.

This equation is not in quadratic form.

Factor the polynomial as if x^2, not x, is the variable: $y = x^4 - 13x^2 + 36 = (x^2 - 9)(x^2 - 4)$

$(x^2 - 9)$ can be factored as $(x - 3)(x + 3)(x^2 - 4)$.

$(x^2 - 4)$ can be factored as $(x - 2)(x + 2)$: $y = (x - 3)(x + 3)(x - 2)(x + 2)$.

The x-intercepts of the polynomial are $x = 2, x = -2, x = 3$, and $x = -3$.

Practice Questions

1. Given the roots of a polynomial are $-8, -5,$ 4, and 12, write a possible equation of the polynomial as a product of its factors.

2. Write a possible equation of the graph below as a product of its factors.

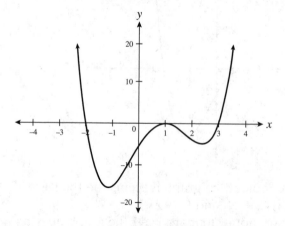

3. Find the x-intercepts of the polynomial $y = x^3 - x^2 - 16x + 16$.

4. Find the zeros of the polynomial $y = x^4 - 5x^2 + 4$.

Practice Answers

1. A possible equation of the polynomial is $y = (x + 8)(x + 5)(x - 4)(x - 12)$.
2. A possible equation of the polynomial is $y = (x + 2)(x - 1)^2 (x - 3)$.
3. The x-intercepts of the polynomial are $x = 1, x = 4$, and $x = -4$.
4. The zeros of the polynomial are $x = 1, x = -1, x = 2$, and $x = -2$.

SOLVED SAT PROBLEMS

1. If the roots of a polynomial are -5, 2, and 8, which of the following choices is a factor of the polynomial?

    A. $x - 5$
    B. $x + 5$
    C. $x + 2$
    D. $x + 8$

    EXPLANATION: Choice B is correct.

    Since -5, 2, and 8 are the roots, $x + 5$, $x - 2$, and $x - 8$ are the factors. Only choice B, $x + 5$, gives one of the correct factors.

2. The function $f(x)$ is graphed in the xy-plane below and $g(x) = f(x) + c$, where c is a constant. Which of the following could be the value of c so that $g(x)$ has four real solutions?

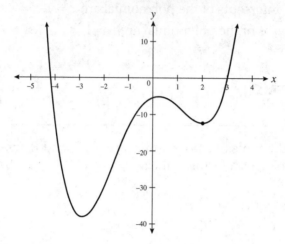

A. −10
B. 0
C. 10
D. 20

EXPLANATION: Choice C is correct.

If $c = 10$, then the entire graph would be shifted up 10 spaces, and the graph of $g(x)$ would be the diagram shown below.

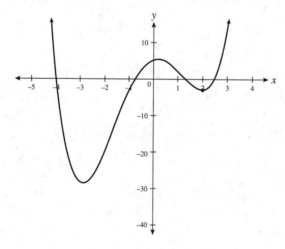

POLYNOMIALS FACTORS AND ROOTS
PRACTICE SAT QUESTIONS

ANSWER SHEET

Choose the correct answer.

1. Ⓐ Ⓑ Ⓒ Ⓓ
2. Ⓐ Ⓑ Ⓒ Ⓓ
3. Ⓐ Ⓑ Ⓒ Ⓓ
4. Ⓐ Ⓑ Ⓒ Ⓓ
5. Ⓐ Ⓑ Ⓒ Ⓓ

PRACTICE SAT QUESTIONS

1. $x^3 - 3x^2 - 3x + 9$

   What irrational value of x is a solution for the equation above?

   A. $\pm\sqrt{3}$
   B. 3
   C. $3i$
   D. $-3i$

2. Which of the following could be the graph of a function with exactly one real root?

   A.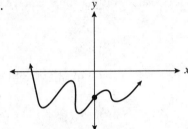

   B.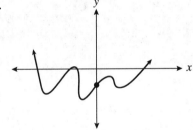

   C.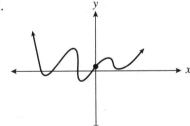

   D.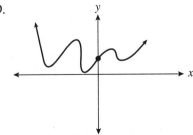

3. Which of the following could be the equation for a function $g(x)$ with zeros at -4, -1, and 7?

   A. $g(x) = (x-4)(x-1)(x+7)$
   B. $g(x) = (x)(x-4)(x-1)(x+7)$
   C. $g(x) = (x+4)(x+1)(x-7)$
   D. $g(x) = (x)(x-4)(x-1)(x+7)$

4. $(x+9)$ is a factor of a polynomial $h(x)$. Which of the following must be true about $h(x)$?

   A. 18 is a root of function $h(x)$.
   B. 9 is a root of function $h(x)$.
   C. -9 is a root of function $h(x)$.
   D. $\dfrac{1}{9}$ is a root of function $h(x)$.

5.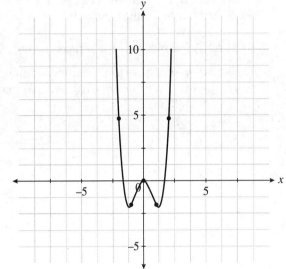

   The graph of $y = x^4 - 2.8x^2$ is shown above. Assume s is a constant and the equation $f(x) = s$ has four solutions. Which number could be s?

   A. -4
   B. -1
   C. 0
   D. 5

▬ EXPLAINED ANSWERS

1. **EXPLANATION: Choice A is correct.**

    Start by factoring. Group the terms and take out common factors: $x^2(x - 3) - 3(x - 3)$, so $(x - 3)(x^2 - 3)$. Set each of these equal to zero and solve for x. $x = 3$, but that is not an irrational value, so $x^2 = 3$, $x = \pm\sqrt{3}$, answer choice A is correct.

2. **EXPLANATION: Choice A is correct.**

    To find the number of real roots, count the number of times the graph of an equation crosses the x-axis. Inspection reveals that the graph in choice A crosses the x-axis exactly once.

3. **EXPLANATION: Choice C is correct.**

    The zeros can be found from the factors by setting each factor equal to 0 and solving for x. Choice C indicates that there would be zeros at −4, −1, and 7.

4. **EXPLANATION: Choice C is correct.**

    −9 is a root of the function because $x + 9 = 0$ by the definition of a root. The others might be, but are not necessarily, roots of the function.

5. **EXPLANATION: Choice B is correct.**

    The equation will have four solutions only if the graph of $f(x) = s$ crosses the graph of the equation at four points. Of the given choices, that only happens at −1.

CHAPTER 16

RATIONAL EXPRESSIONS, EQUATIONS, AND SYNTHETIC DIVISION

A SAT question may ask you to simply solve a rational equation given a value for x. Others will ask you to determine when a rational expression or equations is **undefined**, meaning it has no solutions. Undefined equations and expressions usually have zero in the denominator. Finally, you will be asked to write an equivalent form of a rational expression or equation. Sometimes this is best done using a tool called **synthetic division**, which will be described in this chapter. In this chapter you will also see how synthetic division can be used to determine factors of polynomials. Note that synthetic division is one possible method, but not the only one, so use whichever method makes you most comfortable. Begin with the mathematics review and then complete and correct the practice problems. There are 2 Solved SAT Problems and 11 Practice SAT Questions with answer explanations.

Solve For a Given Value

Example 1.

$$f(x) = \frac{2x^3 - x^2 + 5x - 1}{x + 5}, \ f(2) = ?$$

$$f(2) = \frac{2(2)^3 - (2)^2 + 5(2) - 1}{(2) + 5} = \frac{2(8) - 4 + 10 - 1}{7} = \frac{16 - 4 + 10 - 1}{7} = \frac{21}{7} = 3$$

Identify Values That Make an Equation Undefined

Example 2.

For what values of x is the function $f(x) = \dfrac{6}{x^2 - 8x + 15}$ undefined?

A function is undefined when the denominator is 0.

Since you cannot divide by 0, find the values of x that make the denominator 0.

Set denominator equal to 0: $x^2 - 8x + 15 = 0$

Factor: $(x - 5)(x - 3) = 0$

Set each factor equal to 0 and solve: $(x - 5) = 0 \rightarrow x = 5$

$(x - 3) = 0 \rightarrow x = 3$

$f(x)$ is undefined when $x = 5$ and $x = 3$.

Example 3.

For what values of x is the function $f(x) = \dfrac{5}{(x+3)^2 - 2(x+3) - 15}$ undefined?

Since you cannot divide by 0, find the values of x that make the denominator 0.

Set the denominator equal to 0: $(x+3)^2 - 2(x+3) - 15$

Factor using $x + 3$ as the first term: $((x+3) - 5)((x+3) + 3) = 0$

Simplify expressions inside each pair of parentheses: $(x-2)(x+6) = 0$

Set each factor equal to 0 and solve: $(x-2) = 0 \rightarrow x = 2$

$\qquad\qquad\qquad\qquad\qquad\qquad\qquad (x+6) = 0 \rightarrow x = -6$

$f(x)$ is undefined when $x = 2$ and $x = -6$.

Factoring When Equations Are Undefined

Example 4.

$f(x) = \dfrac{x^2 - 3x - 28}{x^2 + 12x + 32}$. $x \neq -4$ and $x \neq -8$. Write an equivalent equation that is defined for $x = -4$.

Factor the numerator and denominator: $f(x) = \dfrac{x^2 - 3x - 28}{x^2 + 12x + 32} = \dfrac{(x-7)(x+4)}{(x+8)(x+4)}$

Reduce the expression: $f(x) = \dfrac{(x-7)(x+4)}{(x+8)(x+4)} = \dfrac{(x-7)\cancel{(x+4)}}{(x+8)\cancel{(x+4)}} = \dfrac{(x-7)}{(x+8)}$

So, for $x \neq -4$, $f(x) = \dfrac{x^2 - 3x - 28}{x^2 + 12x + 32} = \dfrac{(x-7)}{(x+8)}$ $x \neq -8$.

The reason the problem states for $x \neq -4$ is because $f(x)$ is undefined when $x = -4$. We can factor to find an equivalent equation because we can factor $(x + 4)$ out of the numerator and the denominator. This means the original equation is not defined for $x = -4$, but the new equation is defined for $x = -4$. It needs to be clear that after simplifying, the new equation is equal to the original equation everywhere except for $x = -4$

Example 5.

$f(x) = \dfrac{x^3 - 4x}{x^2 + 6x - 16}$. $x \neq 2$ and $x \neq -8$. Write an equivalent equation that is defined for $x = 2$.

Factor the numerator and denominator: $f(x) = \dfrac{x^3 - 4x}{x^2 + 6x - 16} = \dfrac{x(x^2 - 4)}{(x-2)(x+8)} = \dfrac{x(x-2)(x+2)}{(x-2)(x+8)}$

Reduce the expression: $f(x) = \dfrac{x(x-2)(x+2)}{(x-2)(x+8)} = \dfrac{x\cancel{(x-2)}(x+2)}{\cancel{(x-2)}(x+8)} = \dfrac{x(x+2)}{(x+8)}$

So, for $x \neq 2$, $f(x) = \dfrac{x^3 - 4x}{x^2 + 6x - 16} = \dfrac{x(x+2)}{(x+8)}$.

The reason the problem states for $x \neq 2$ is because $f(x)$ is undefined when $x = 2$. We can factor to find an equivalent equation because we can factor

$(x - 2)$ out of the numerator and the denominator. This means the original equation is not defined for $x = 2$, but the new equation is defined for $x = 2$. It needs to be clear that after simplifying, the new equation is equal to the original equation everywhere except for the undefined value $x = 2$.

Example 6.

Write an equation equivalent to $f(x) = \dfrac{2}{x - 3} + \dfrac{x + 4}{x^2 - 9}$. $x \neq 3$ and $x \neq -3$.

Factor the denominator of the second fraction: $f(x) = \dfrac{2}{x - 3} + \dfrac{x + 4}{x^2 - 9} = \dfrac{2}{x - 3} +$
$\dfrac{x + 4}{(x - 3)(x + 3)}$

Multiply the first fraction by $\dfrac{x + 3}{x + 3}$ to create the common denominator:

$$f(x) = \left(\dfrac{x + 3}{x + 3}\right)\dfrac{2}{x - 3} + \dfrac{x + 4}{(x - 3)(x + 3)}$$

Rewrite first fractions: $f(x) = \dfrac{2x + 6}{(x - 3)(x + 3)} + \dfrac{x + 4}{(x - 3)(x + 3)}$

Both fractions have the same denominator, so add and simplify:

$$f(x) = \dfrac{3x + 10}{(x - 3)(x + 3)} = \dfrac{3x + 10}{x^2 - 9}. \; x \neq 3 \text{ and } x \neq -3. \text{ (Notice that this equation}$$

is still undefined for $x = 3$ and $x = -3$.)

Synthetic Division

Example 7.

Write an expression equivalent to $\dfrac{x^3 - 4x^2 - 11x + 30}{x - 5}$. (As with this example, the expression has to be in descending order with no missing terms.)

Synthetic division can be useful when you are dividing by a linear expression, an expression in the form $ax + b$. There is really no sense to it—just follow the steps.

Set the linear expression in the denominator equal to 0 and solve:
$x - 5 = 0 \rightarrow x = 5$.

Now set up the synthetic division problem where the outside number is the value of x, in this case 5, and the inside numbers are the coefficients of the expression in the numerator.

$$5 | 1 \quad -4 \quad -11 \quad 30$$

Then bring the leading coefficient (first number) inside the division box straight down, multiply the number outside division box, 5, with the number you brought down, put the result in the next column, add the two numbers together, and write the result, $-4 + 5 = 1$ as shown below.

$$
\begin{array}{r|rrrr}
5 & 1 & -4 & -11 & 30 \\
  &   & 5 &   &   \\
\hline
  & 1 & 1 &   &   \\
\end{array}
$$

Multiply the number outside division box, 5, with the result of the addition, 1, from the column, put the result, 5, in the next column, add $-11 + 5$, and write the result $-11 + 5 = -6$ as shown below.

$$
\begin{array}{r|rrrr}
5 & 1 & -4 & -11 & 30 \\
  &   & 5 & 5   &   \\
\hline
  & 1 & 1 & -6 &
\end{array}
$$

Repeat the process and write the result as shown below.

$$
\begin{array}{r|rrrr}
5 & 1 & -4 & -11 & 30 \\
  &   & 5 & 5   & -30 \\
\hline
  & 1 & 1 & -6 & 0
\end{array}
$$

The numbers 1, 1, −6 in the bottom row of the synthetic division problem are the coefficients of the solution starting with x^2, the exponent one less than the original x^3, while the final number, 0, is the remainder. Therefore $\dfrac{x^3 - 4x^2 - 11x + 30}{x - 5} = 1x^2 + 1x - 6 + \dfrac{0}{x - 5} = x^2 + x - 6.$

Since the final number in the bottom row is 0, the linear expression in the denominator divides evenly into the numerator, the denominator is a factor of the numerator, and $x = 5$ is a root of the numerator. This can be confirmed by substituting $x = 5$ into the numerator and solving:

$$(5)^3 - 4(5)^2 - 11(5) + 30 = 125 - 4(25) - 55 + 30 = 125 - 100 - 55 + 30 = 0.$$

Example 8.

Write an expression equivalent to $\dfrac{x^3 + 3x^2 - 6x - 8}{2x + 4}$.

Let's use *synthetic division*

Set the linear expression in the denominator equal to 0 and solve: $2x + 4 = 0 \rightarrow x = -2.$

Now set up the synthetic division problem where the outside number where the outside number is the value of x, in this case −2, and the inside numbers are the coefficients of the expression in the numerator.

$$
\begin{array}{r|rrrr}
-2 & 1 & 3 & -6 & -8
\end{array}
$$

Then bring the leading coefficient (first number) straight down, multiply the number outside division box, −2, with the number you brought down, put the result in the next column, add the two numbers together, and write the result $3 + -2 = 1$, as shown below

$$
\begin{array}{r|rrrr}
-2 & 1 & 3 & -6 & -8 \\
   &   & -2 &   &   \\
\hline
   & 1 & 1 &   &
\end{array}
$$

Multiply the number outside the division box, –2, with the result of the addition, 1, put the result, –2, in the next column, add $-6 + -2 = 8$, and write the result as shown below.

$$
\begin{array}{r|rrrr}
-2 & 1 & 3 & -6 & -8 \\
   &   & -2 & -2 &   \\
\hline
   & 1 & 1 & -8 &   \\
\end{array}
$$

Repeat the process and write the results as shown below.

$$
\begin{array}{r|rrrr}
-2 & 1 & 3 & -6 & -8 \\
   &   & -2 & -2 & 16 \\
\hline
   & 1 & 1 & -8 & 8 \\
\end{array}
$$

The numbers 1, 1, –8 in the bottom row of the synthetic division problem are the coefficients of the solution starting with x^2, the exponent one less than the original x^3, while the final number, 8, is the remainder. Therefore,

$$\frac{x^3 + 3x^2 - 6x - 8}{2x + 4} = 1x^2 + 1x - 8 + \frac{8}{2x + 4} = x^2 + x - 8 + \frac{4}{x + 2}.$$

Since the final number in the bottom row is not 0, the linear expression in the denominator does not divide evenly into the numerator. It is not a factor of the numerator and $x = -2$ is not a root of the numerator. However, the remainder is the result of evaluating the numerator at $x = -2$. This can be confirmed by substituting $x = -2$ into numerator and solving:

$$(-2)^3 + 3(-2)^2 - 6(-2) - 8 = -8 + 3(4) + 12 - 8 = -8 + 12 + 12 + -8 = 8.$$

Practice Questions

1. For what values of x is the function $f(x) = \dfrac{6}{x^2 + 4x - 21}$ undefined?

2. For what values of x is the function $f(x) = \dfrac{5}{(x - 7)^2 + 4(x - 7) - 12}$ undefined?

3. Write an equation equivalent to $f(x) = \dfrac{x^2 - 9x + 18}{x^2 - 2x - 24}$ for $x \neq 6$.

4. Write an equation equivalent to $f(x) = \dfrac{5}{x + 6} + \dfrac{2x - 7}{x^2 + 12x + 36}$.

5. Write an expression equivalent to $\dfrac{x^3 - 6x^2 + 3x + 10}{x - 3}$.

Practice Answers

1. $x = 3$ and $x = -7$.
2. $x = 1$ and $x = 9$.
3. $f(x) = \dfrac{x - 3}{x + 4}$, $x \neq -4$.
4. $f(x) = \dfrac{7x + 23}{(x + 6)^2}$, $x \neq -6$.
5. $x^2 - 3x - 6 - \dfrac{8}{x - 3}$.

1. Which of the following is equivalent to $\dfrac{3x^2 + x - 10}{3x - 6}$?

   A. $3x - 5$

   B. $3x + 7$

   C. $3x - 5 + \dfrac{4}{3x - 6}$

   D. $3x + 7 + \dfrac{4}{3x - 6}$

EXPLANATION: Choice D is correct.

Set the linear expression in the denominator equal to 0 and solve:

$$3x - 6 = 0 \rightarrow x = 2.$$

Now set up the synthetic division problem where the outside number is what made the linear expression in the denominator 0, in this case 2, and the inside numbers are the coefficients of the expression in the numerator.

$$2 \underline{\smash{| \quad 3 \quad 1 \quad -10}}$$

Once the problem is set up correctly, bring the leading coefficient (first number) straight down, multiply the number outside division box, 2, with the number you brought down, put the result in the next column, add the two numbers together, and write the result in the bottom of the row.

$$\begin{array}{r|rrr} 2 & 3 & 1 & -10 \\ & & 6 & \\ \hline & 3 & 7 & \end{array}$$

Multiply the number outside the division box, 2, with the result of the addition from the second column, put the result in the next column, add the two numbers together, and write the result in the bottom of the row.

$$\begin{array}{r|rrr} 2 & 3 & 1 & -10 \\ & & 6 & 14 \\ \hline & 3 & 7 & 4 \end{array}$$

The coefficients of the equivalent expression to $\dfrac{3x^2 + x - 10}{3x - 6}$ are made up of the numbers in the bottom row of the synthetic division and the final number in the row is the remainder of the division. The values of the exponents start off one value less than the original numerator and go down one with each term. Therefore, $\dfrac{3x^2 + x - 10}{3x - 6} = 3x + 7 + \dfrac{4}{3x - 6}$.

2. When the polynomial $f(x)$ is divided by $x + 8$, the remainder is 0.
Which of the following statements must be true?

   A. $f(8) = 0$

   B. $f(0) = -8$

   C. $x = -8$ is a root of $f(x)$.

   D. $x - 8$ is a factor of $f(x)$.

   EXPLANATION: Choice C is correct.

   Since $f(x)$ is divided by $x + 8$ has a remainder is 0, $x + 8$ is a factor of $f(x)$, so $x = -8$ is a root of $f(x)$.

RATIONAL EXPRESSIONS, EQUATIONS, AND SYNTHETIC DIVISION PRACTICE SAT QUESTIONS

ANSWER SHEET

Choose the correct answer.
If no choices are given, grid the answers in the section at the bottom of the page.

1. Ⓐ Ⓑ Ⓒ Ⓓ
2. Ⓐ Ⓑ Ⓒ Ⓓ
3. Ⓐ Ⓑ Ⓒ Ⓓ
4. Ⓐ Ⓑ Ⓒ Ⓓ
5. Ⓐ Ⓑ Ⓒ Ⓓ
6. Ⓐ Ⓑ Ⓒ Ⓓ
7. Ⓐ Ⓑ Ⓒ Ⓓ
8. GRID
9. GRID
10. Ⓐ Ⓑ Ⓒ Ⓓ

11. Ⓐ Ⓑ Ⓒ Ⓓ

Use the answer spaces in the grids below if the question requires a grid-in response.

| Student-Produced Responses | ONLY ANSWERS ENTERED IN THE CIRCLES IN EACH GRID WILL BE SCORED. YOU WILL NOT RECEIVE CREDIT FOR ANYTHING WRITTEN IN THE BOXES ABOVE THE CIRCLES. |

8.

9.

PRACTICE SAT QUESTIONS

1. What is the domain of $f(x) = \dfrac{4}{x^2 + x - 6}$?

   A. All real numbers except 4
   B. All real numbers except 0
   C. All real numbers except 6
   D. All real numbers except −3 and 2

2. Which of the following is equivalent to the sum of the expressions $b^2 - 3$ and $b + 3$?

   A. $b^3 - 1$
   B. $2b^2$
   C. b^3
   D. $b^2 + b$

3. For which value of x in the expression below is the expression undefined?

   $$\dfrac{16}{x^2 - 4x + 3}$$

   A. −2
   B. −1
   C. 0
   D. 1

4. For which value of x is the equation below undefined?

   $$(x + 3)^2 + (2x - 2) - 5x - 2 = 3$$

   A. −4
   B. −2
   C. 2
   D. 3

5. There are two functions $h(x)$ and $j(x)$:

   $$h(x) = x^2 + 3x - 18$$
   $$j(x) = x^2 + 4x - 12$$

   Which of the following is equivalent to $\dfrac{h(x)}{j(x)}$?

   A. The function is undefined.
   B. 1
   C. $\dfrac{x - 2}{x - 3}$
   D. $\dfrac{x - 3}{x - 2}$

6. When $4 < x < 10$, which of the following expressions would be another way to write

   $$\dfrac{1}{\dfrac{2}{x + 3} + \dfrac{2}{x + 5}}?$$

   A. $x^2 - 7$
   B. $x^2 + 5x - 10$
   C. $\dfrac{x^2 - 7}{x^2 + 5x + 10}$
   D. $\dfrac{x^2 + 8x + 15}{4x + 16}$

7. Which of the following is a possible value that satisfies the equation $5x^2 + 6x + 1 = 0$?

   A. -6 ± 4
   B. -3 ± 2
   C. $\dfrac{-6 \pm 4}{10}$
   D. $\dfrac{-3 \pm 2}{10}$

8. $x(x^4 - 13x^2) = -36x$

   What is one possible positive integer solution to the equation above?

9. $(-5x^2 + 13x - 4) - 7(x^2 - 8x - 10)$

   If this expression is rewritten in standard $ax^2 + bx + c$ form, what is the value of the sum of b and c?

10. The point (6, 2) is a valid solution to the function $h(x)$. If $h(x) = d + 3x$ and d is assumed to be a constant, find d.

    A. −16
    B. 0
    C. 6
    D. 16

11. $3nx - 20 = 4(x - 4) + 2(x + 6)$

    The given equation has no solutions. What is the value of n?

    A. −2
    B. 0
    C. 2
    D. 6

EXPLAINED ANSWERS

1. **EXPLANATION: Choice D is correct.**

    Write the denominator equal to 0 to find the values that are not part of the domain. This technique works because the value of the denominator can never equal 0.

    $$f(x) = \frac{4}{x^2 + x - 6}$$

    so write $x^2 + x - 6 = 0$

    $(x + 3)(x - 2) = 0 \Rightarrow x = -3$ and $x = 2$.

    The values (-3) and (2) cannot be part of the domain.

2. **EXPLANATION: Choice D is correct.**

    Many problems with exponential expressions like this one can be solved by plugging in numbers. For example, make $b = 4$. The first expression then becomes $16 - 3 = 13$, and the second becomes $4 + 3 = 7$. Add the two together: 20 is the answer you're looking for. Now plug 4 into each choice to see what you get. Choice A gives you 63—eliminate it. Choice B gives you 32, which is still too high. Choice C gives you 64. Choice D gives you $16 + 4 = 20$, the correct answer. Note that you can plug in any number that follows the rules of the problem, but always check all four answer choices!

3. **EXPLANATION: Choice D is correct.**

    The expression is undefined when $x^2 - 4x + 3 = 0$. You can either work backwards from the answers, or you can solve the equation algebraically. Plugging in 1, answer choice D, gives you: $1^2 - 4(1) + 3$, which simplifies to $1 - 4 + 3$. This equals 0, making the expression undefined.

4. **EXPLANATION: Choice B is correct.**

    The expression is undefined when the value of the denominator is zero. Start by using FOIL to discover that $(x + 3)^2$ becomes $x^2 + 6x + 9$. Now you should have $x^2 + 6x + 9 + 2x - 2 - 5x - 2 = 3$. Combine like terms: $x^2 + 3x + 5 = 3$ or $x^2 + 3x + 2 = 0$. Now factor: $(x + 1)(x + 2)$: the expression is undefined when $x = -1$ or $x = -2$. Only -2 is given as an answer choice, so that is the credited response.

5. **EXPLANATION: Choice D is correct.**

    Start by factoring the functions on the top and on the bottom: $h(x)$ becomes $(x - 3)(x + 6)$ and $j(x)$ becomes $(x - 2)(x + 6)$. Cancel the $(x + 6)$ term, leaving $\frac{x - 3}{x - 2}$.

6. **EXPLANATION: Choice D is correct.**

    Start by plugging in a number for x: the question specifies that it must be between 4 and 10, so let's use 5. The bottom half of the equation then becomes $\frac{2}{8} + \frac{2}{10}$. Put the fractions in the common denominator. Now your equation looks like $\frac{1}{\frac{10}{40} + \frac{8}{40}}$ or $\frac{1}{\frac{18}{40}}$. To simplify the fraction, multiply by the reciprocal, which gives you $\frac{40}{18}$. Now look to the answer choices and plug in 5 for x to find an answer that is equivalent. Choice A gives you 18, the fraction's denominator, and choice B gives you 40, the numerator. Look for a fraction expression that combines the two correctly: choice D ($\frac{80}{36}$, which is equivalent to $\frac{40}{18}$).

7. **EXPLANATION: Choice C is correct.**

Use the quadratic formula, $x = \dfrac{-b \pm \sqrt{b^2 - 4ac}}{2a}$, to solve. b in this case is 6, a is 5, and c is 1. Knowing the quadratic formula off the top of your head will likely allow you to easily answer at least one question per test that might otherwise be difficult or impossible, since sometimes the quadratics on the test cannot be solved by factoring.

8. **EXPLANATION: The correct answer is either 3 or 2.**

Start by putting the equation in standard form and factoring out an x: $x(x^4 - 13x^2 + 36) = 0$. Now factor the equation: it becomes $(x^2 - 4)(x^2 - 9)$. Each of those pieces can then be factored further: $(x - 2)(x + 2)(x - 3)(x + 3)$. Thus, $x = 2, -2, -3,$ or 3. You can grid in either of the positive answers, 3 or 2, to get credit for the question.

9. **EXPLANATION: The correct answer is 135.**

Begin by distributing -7, so that the given equation becomes $(-5x^2 + 13x - 4) - 7x^2 + 56x + 70$. Then, combine like terms: $-12x^2 + 69x + 66$. Now add b (69) and c (66) to get 135, the answer.

10. **EXPLANATION: Choice A is correct.**

Because you have a point that is a solution to the function, you can plug that point in to solve for d, using the y-coordinate to represent $h(x)$ and the x-coordinate as x. $2 = d + 3(6)$. Simplify: $2 = d + 18$, and $d = -16$. The other answers stem from sign mistakes or from putting the point into the equation backwards.

11. **EXPLANATION: Choice C is correct.**

First, distribute to expand the right side of the equation: $3nx - 20 = 4x - 16 + 2x + 12$. Simplify: $3nx - 20 = 6x - 4$. The equation will have no solutions if the coefficients of x are the same on each side but the numbers being added or subtracted are different. Therefore, we need $3nx$ to be the same as $6x$, which means n must be 2 in order to give the equation no solutions.

CHAPTER 17

EXPONENTIAL FUNCTIONS

Many of the SAT questions about **exponential functions** focus on percent increase and percent decrease. You will also match tables and graphs to a specific exponential function. Begin with the mathematics review and then complete and correct the practice problems. There are 2 Solved SAT Problems and 15 Practice SAT Questions with answer explanations. The general formulas for increase and decrease are:

Increase: $f(x) = a(1 + r)^t$

Decrease: $f(x) = a(1 - r)^t$

[a = initial amount, r = growth or decay rate, t = number of time interval.]

Example 1.

Quinn deposits $500 in a savings account that earns 3% interest per year compounded annually. Write a function f that models the amount of money in the account after t years. Then, use the function to find how much is in the account after 7 years rounded to the nearest hundredth.

The function is $f(t) = 500(1 + 0.03)^t$ $\Rightarrow$ $f(t) = 500(1.03)^t$, where t is the number of years.

Substitute $t = 7$: $f(7) = 500(1.03)^7 \approx \$614.94$

Example 2.

There are 65 grams of a radioactive substance, and the substance decays at an annual rate of 4% per year. Write a function f that models the number of grams of the substance remaining after t years. Then use the function to find the number of grams of the substance remaining after 12 years rounded to the nearest hundredth.

The function is $f(t) = 65(1 - 0.04)^t$ $\Rightarrow$ $f(t) = 65(0.96)^t$, where t is the number of years.

Substitute $t = 12$: $f(12) = 65(0.96)^{12} \approx 39.83$ grams.

Example 3.

On the first day of a new music app being available for download, 120 people signed up to use the app. After that, it had an increase of 9% of users per week. Write a function f that models the number of people who signed up for the music app after w weeks. Then use the function to find the number of people who have signed up after 10 weeks.

The function is $f(w) = 120(1 + 0.09)^w$ $\Rightarrow$ $f(w) = 120(1.09)^w$, where w is the number of weeks.

Substitute $w = 10$: $f(10) = 120(1.09)^{10} \approx 284$ people.

Example 4.

A doctor administers 400 milligrams of a medicine to a patient. The amount of medicine in the patient's system decays by 2.1% every 10 minutes. Write a function f that models the number of milligrams of medicine in the patient's system after h hours. Then use the function to find the number of milligrams of medicine in the patient's system after 3 hours to the nearest milligram.

   The first thing to deal with is that the decay rate is given per 10 minutes, but the equation is to be written per hour. There are 60 minutes in an hour, so every hour there will be 6 instances of the 2.1% decay rate. Therefore, every hour(h) there are $6h$ cases of the 2.1% decay rate.

   The function is $f(h) = 400(1 - 0.021)^{6h}$ $\Rightarrow$ $f(h) = 400(0.979)^{6h}$, where h is the number of hours.

   Substitute $h = 3$: $f(3) = 400(0.979)^{6(3)} = 400(0.979)^{18} \approx 273$ milligrams.

Example 5.

There are 40 grams of a radioactive substance, and the substance has a half-life of 15 hours. Write a function f that models the number of grams of the substance remaining after h hours. Then use the function to find the number of grams of the substance remaining after 3 days rounded to the nearest hundredth.

   The decay rate is given per 15 hours, but the equation is to be written per hour. After 15 hours, the amount decays by half. After 30 hours, the amount decays by half a second time, and after 45 hours, half again a third time, and so on. So, the amount will decay by half every $\dfrac{h}{15}$ hours.

   The function is $f(h) = 40(0.5)^{\frac{h}{15}}$, where h is the number of hours.

   There are 24 hours in a day, so $3(24) = 72$ hours in 3 days.

   Substitute $h = 72$: $f(72) = 40(0.5)^{\frac{72}{15}} = 40(0.5)^{4.8} \approx 1.44$ grams.

Example 6.

Compute the function value for $f(x) = 2^x$ at $x = -2, -1, 0, 1,$ and 2. Then draw a graph that models the function $f(x) = 2^x$.

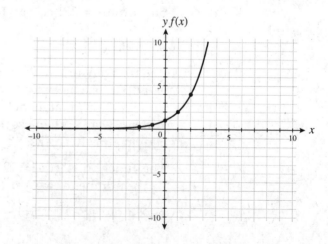

x	$f(x)$ y
-2	$2^{-2} = \dfrac{1}{2^2} = \dfrac{1}{4}$
-1	$2^{-1}\dfrac{1}{2^1} = \dfrac{1}{2}$
0	$2^0 = 1$
1	$2^1 = 2$
2	$2^2 = 4$

Note: The graph never touches the x-axis.

Example 7.

Draw the graph of $g(x) = 2^x + 3$.

The graph of $g(x) = 2^x + 3$ can be created by shifting the graph of $f(x) = 2^x$ up 3 units.

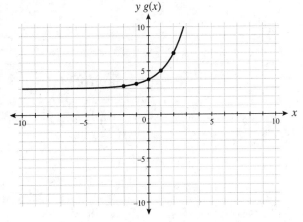

Note: The graph never touches $y = 3$.

Example 8.

Draw the graph of $h(x) = -2^x$.

The graph of $h(x) = -2^x$ can be created by flipping the graph of $f(x) = 2^x$ over the x-axis.

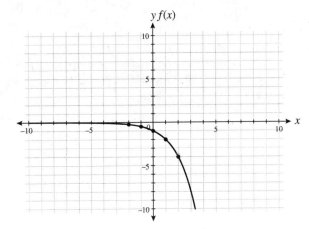

Note: The graph never touches the x-axis.

Example 9.

Compute the function value for $f(x) = 3^{-x}$ at $x = -2, -1, 0, 1,$ and 2. Then draw a graph that models the function $f(x) = 3^{-x}$.

x	$f(x)$ y
-2	$3^{-(-2)} = 3^2 = 9$
-1	$3^{-(-1)} = 3^1 = 3$
0	$3^0 = 1$
1	$3^{-(1)} = \dfrac{1}{3^1} = \dfrac{1}{3}$
2	$3^{-(2)} \dfrac{1}{3^2} = \dfrac{1}{9}$

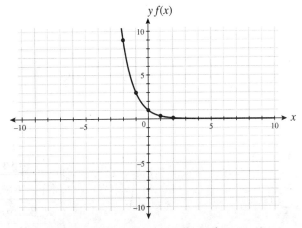

Note: The graph never touches the x-axis.

Example 10.

Draw the graph of $g(x) = 3^{-x} - 2$.

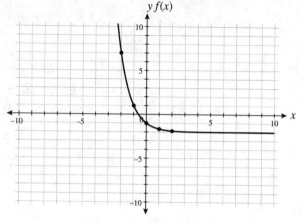

The graph of $g(x) = 3^{-x} - 2$ can be created by shifting the graph of $f(x) = 3^{-x}$ down 2 units.

Note: The graph never touches $y = -2$

Example 11.

Draw the graph of $h(x) = -3^{-x} + 1$.

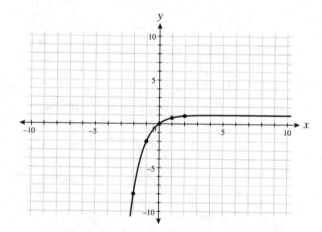

The graph of $h(x) = -3^{-x} + 1$ can be created by flipping the graph of $f(x) = 3^{-x}$ over the x-axis, then shifting the graph up 1 unit.

Note: The graph never touches $y = 1$.

Some examples of growth and decay are not strictly exponential. For example, geometric growth is modeled by a function with an exponential value multiplied by a constant.

Example 12.

In an experiment, there were initially 5 bacteria. The number of bacteria doubles every hour. Write an equation and draw a graph that models the number of bacteria, y, based on the number of hours (x) that have passed.

Time Passed	Number of Bacteria
0 hours	$5 \times (2^0) = 5$
1 hour	$5 \cdot 5(2^1) = 10$
2 hours	$5 \cdot 2^2 = 20$
3 hours	$5 \cdot 2^3 = 40$
4 hours	$5 \cdot 2^4 = 80$

After each hour, multiply the previous value by 2 to get the next value, $y = f(x) = 5 \cdot 2^x$.

This is the geometric growth model where $y = 5(2^x)$. The hour 0 through hour 4, the y values are 5, 10, 20, 40, 80.

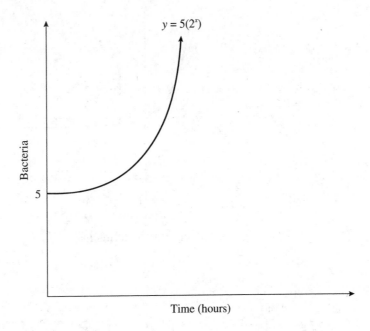

Practice Problems

1. Robert invests $800 into a stock that he believes will increase by 5% each month for the next year. Assuming Robert is correct, write a function f that models the value of Robert's investment over the next year after m months. Then use the function to find how much is in the account after 9 months rounded to the nearest hundredth.

2. A certain discount warehouse store discounts each item in the store by 6% a week. Erin notices a dresser that she likes, and the original price of the item is $150. Write a function f that models the price of the dresser after w weeks. Then use the function to find out how much Erin will spend of the dresser if she waits 4 weeks to buy it. Round your answer to the nearest hundredth.

3. There are 115 grams of a radioactive substance, and the substance has a half-life of 8 days. Write a function f that models the number of grams of the substance remaining after d days. Then use the function to find the number of grams of the substance remaining after 35 days rounded to the nearest hundredth.

4. Draw the graph of $f(x) = 2^{-x} - 1$.

5. Draw the graph of $g(x) = -3^x + 2$.

Practice Answers

1. $f(m) = 800(1 + 0.05)^m \quad \Rightarrow \quad f(m) = 800(1.05)^m$

    $f(9) = 800(1.05)^9 \approx \$121.06.$

2. $f(w) = 150(1 - 0.06)^w \quad \Rightarrow \quad f(w) = 150(0.94)^w$

    $f(4) = 150(0.94)^4 \approx \$117.11.$

3. $f(d) = 115(0.5)^{\frac{d}{8}}$.

$f(35) = 115(0.5)^{\frac{35}{8}} = 115(0.5)^{4.375} \approx 5.54$ grams.

4.

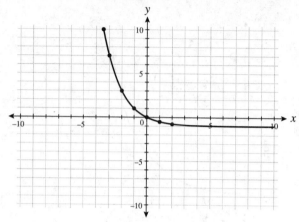

Note: The graph never touches $y = -1$.

5.

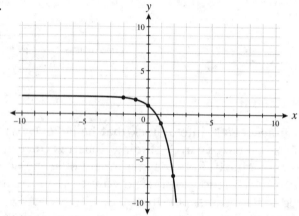

Note: The graph never touches $y = 2$

SOLVED SAT PROBLEMS

1. The equation below models the amount of money, A, in a bank account after t years, where interest is compounded annually.

$$A = 2{,}000(1.03)^t$$

Which of the following choices models the amount of money in the bank account after m months?

A. $A = 2{,}000(1.0025)^{\frac{m}{12}}$

B. $A = 2{,}000(1.0025)^{12m}$

C. $A = 2{,}000(1.03)^{\frac{m}{12}}$

D. $A = 2{,}000(1.03)^{12m}$

EXPLANATION: Choice C is correct.

The only part of the equation that is changing is the exponent, not the base. This eliminates choice A and choice B. The correct exponent is the one that properly converts months into years. Because 12 months is equal to 1 year, the correct choice is the one that gives a result of 1 when 12 is substituted in for m.

Choice C $\dfrac{m}{12}$: $\dfrac{(12)}{12} = 1$ Correct

Choice D $12m$: $12(12) = 144$ Incorrect.

2. Let $f(x) = 3^x - 2$. Which of the following choices is the graph of
    $y = -f(x)$?

A.
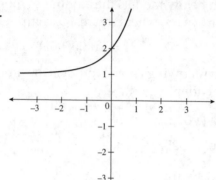

B.
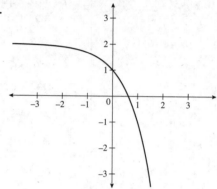

C.
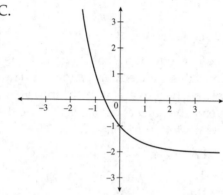

D.
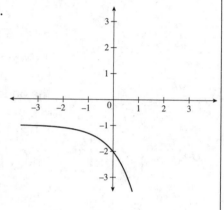

EXPLANATION: Choice B is correct.

The graph of $y = -f(x)$ will be the graph of $f(x) = 3^x - 2$ flipped
across the x-axis.

First graph $y = 3^x - 2$, shown in the solid line graph below. Then flip
it over the x-axis. (Multiplying a function by -1 flips the graph of the
function over the x-axis, as shown in the dashed-line graph below)

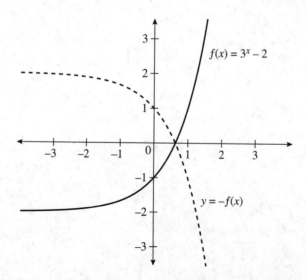

EXPONENTIAL FUNCTIONS
PRACTICE SAT QUESTIONS

ANSWER SHEET

Choose the correct answer.
If no choices are given, grid the answers in the section at the bottom of the page.

1. Ⓐ Ⓑ Ⓒ Ⓓ
2. GRID
3. GRID
4. GRID
5. Ⓐ Ⓑ Ⓒ Ⓓ
6. Ⓐ Ⓑ Ⓒ Ⓓ
7. GRID
8. Ⓐ Ⓑ Ⓒ Ⓓ
9. Ⓐ Ⓑ Ⓒ Ⓓ
10. Ⓐ Ⓑ Ⓒ Ⓓ

11. Ⓐ Ⓑ Ⓒ Ⓓ
12. Ⓐ Ⓑ Ⓒ Ⓓ
13. Ⓐ Ⓑ Ⓒ Ⓓ
14. Ⓐ Ⓑ Ⓒ Ⓓ
15. Ⓐ Ⓑ Ⓒ Ⓓ

Use the answer spaces in the grids below if the question requires a grid-in response.

| Student-Produced Responses | ONLY ANSWERS ENTERED IN THE CIRCLES IN EACH GRID WILL BE SCORED. YOU WILL NOT RECEIVE CREDIT FOR ANYTHING WRITTEN IN THE BOXES ABOVE THE CIRCLES. |

2.

3.

4.

7.

PRACTICE SAT QUESTIONS

Use the table and graph below to answer questions 1 and 2.

A millibar is a way to measure atmospheric pressure. We also use millibars to measure the vapor pressure to find the amount of water vapor in the atmosphere. Higher vapor pressure means more water vapor. The saturation vapor pressure is the amount of water vapor that can be held in the air at any given temperature. As the table and graph below show, saturation vapor pressure increases with temperature.

Temperature Celsius	Approximate Saturation Vapor Pressure (millibars)
0	5
10	9
20	18
30	32

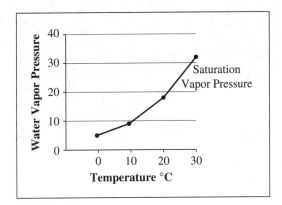

1. Centibars can also be used to measure saturation vapor pressure. There are 10 millibars in a centibar. What is the difference, in centibars, between the approximate saturation vapor pressure at zero degrees Celsius and the approximate saturation vapor pressure at 30 degrees Celsius?

   A. 0.5
   B. 2.7
   C. 3.2
   D. 3.7

2. The formula below is used to approximate saturation vapor pressure (x) given temperature in degrees Fahrenheit (T).

   $(0.0041(\mathbf{T}) + 0.676)^8 - 0.000019\,(\mathbf{T} + 16) + 0.001316$

   Using the formula, what is the saturation vapor pressure to the nearest hundredth if the temperature is 100°F?

3. A town hired a consultant to help determine short-term and long-term population growth. The consultant uses this formula:

   Final population (p) = Starting population (s)
   $(1 + $ Growth rate $(g))$ period in years (y)
   $p = s\,(1 + g)^{(y)}$

   The town currently has 5,000 inhabitants. The consultant estimates the yearly growth rate at 6% a year, and the consultant will project the growth to find p. What value should the consultant use for g?

4. Rounded to 100 people, at a 6% growth rate, how many inhabitants will there be after 4 years?

5. Construction on a high-rise building doubled the number of floors built each year until after six years there were a total of 189 floors. How many floors were built during the third year?

   A. 3
   B. 6
   C. 12
   D. 96

6. The Friendly Savings Bank offered two long-term CD rates. The Super Saver CD rate offered a 3.5% annual rate, while the Premier CD rate was 3.75% annually. Friendly Savings Bank uses the formula (D)eposit $(1 + r)^t$ to find the value of a CD after t years, in which the variable (D) is the initial deposit in dollars.

   What number is substituted for r for the Premier CD?

   A. 0.0375
   B. 0.375
   C. 1.0375
   D. 1.375

7. On the same day, a Friendly Savings Bank customer deposited $1,000 in the Super Saver CD and $1,000 in the Premier CD. After exactly five years, what is the combined balance of the two CDs? Round to the nearest dollar. Disregard the dollar sign when gridding in your answer.

8. An Inventory Control manager in a company plans to increase the value of the company's inventory by 2% every five years. The current value of the company's inventory is $750,000. Which of the following expressions shows the value of the company's inventory after t years?

    A. $750,000 (0.02)^{t/5}$
    B. $750,000 (0.02)^{5t}$
    C. $750,000 (1.02)^{t/5}$
    D. $750,000 (1.02)^{5}$

9. Given below are four descriptions of the rate of change in the depth of a lake. Which of the descriptions reflects an exponential change in the lake's depth?

    A. Each month the lake's depth decreases by 0.25% of the current depth.
    B. Each year the lake's depth decreases by some constant amount.
    C. Each month the lake's depth increases by 0.25% of the lake's original depth.
    D. Each year the lake's depth decreases by 1 foot until the lake is dry.

10. A computer programmer wants to decrease the number of lines of code to make the computer program more compact. As the programmer works, the number of lines in the program decreases 4% each year. Which of the following graphs could model that decrease?

    A.
    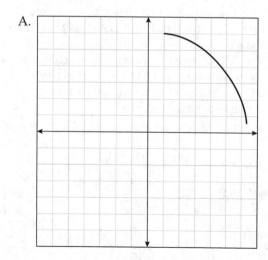

    B.
    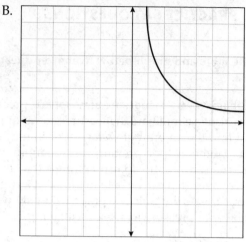

    C.
    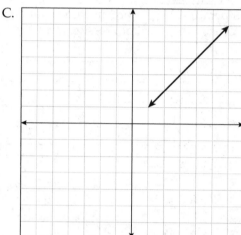

    D.
    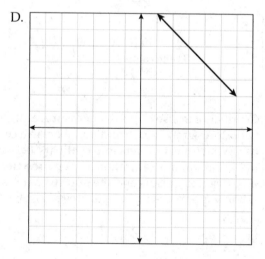

11. A major stockholder in a company sells 2.4% of the number shares held in the company each year, and never adds additional shares. At the beginning of the accounting period, the stockholder has 750,000 shares. Which of the following shows the number of shares remaining after t years?

    A. $750{,}000\,(0.976)^t$
    B. $750{,}000\,(0.024)^t$
    C. $0.976\,(750{,}000)^t$
    D. $0.024\,(750{,}000)^t$

12. The function below defines the value of x in a graph on the x-y plane $g(x) = 2^x - 1$. Which of the graphs below correctly shows $g(x)$?

    A.
    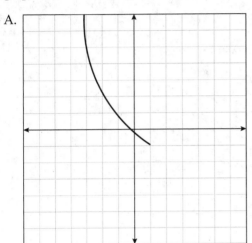

    B.
    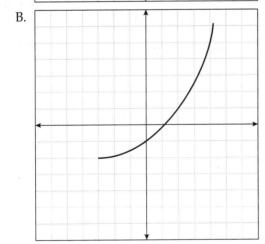

C.
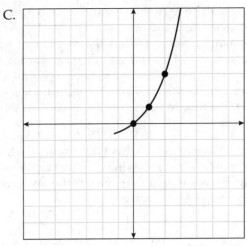

D.
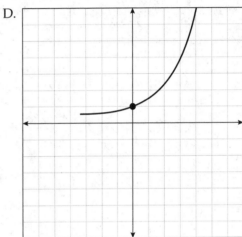

13. The table below shows the depth of water in a pool during a 4-hour cleaning cycle.

Hour	Depth of Water in Feet
1	0.5
2	0.25
3	0.125
4	0.0615

Which of the following is the best way to describe the relationship between the number of hours and the depth of water?

A. It is an example of linear decay.
B. It is an example of exponential decay.
C. It is an example of linear growth.
D. It is an example of exponential growth.

14. A small grocery store features gallon containers of a certain brand of artisanal mineral water. The equation below models the number of gallon containers of water sold in t years starting with the store's grand opening, where G is the number of gallons sold, I is the initial year's sales, and t is the time in years.

$$G = I(2.3)^t$$

If 9,000 gallons are sold in the first year, which of the following models the number of gallons sold in m months after the store's grand opening?

A. $G = (9{,}000 \times 2.3)^{m/12}$
B. $G = (9{,}000 \times 2.3)^{12m}$
C. $G = 9{,}000 \, (2.3)^{m/12}$
D. $G = 9{,}000 \, (2.3)^{12m}$

15. In one small country the annual gross domestic product growth for the past 30 years has been 4.2%. This year the gross domestic product was six million dollars. Assuming it will remain the same this year, the gross domestic product gowth can be represented by which of the following?

A. $(6 \times 4.2)^t$
B. $(6 \times 1.042)^t$
C. $6(4.2)^t$
D. $6(1.042)^t$

EXPLAINED ANSWERS

1. **EXPLANATION: Choice B is correct.**

   Start by converting millibars to centibars for 0 degrees (0.5 centibar) and 30 degrees (3.2 centibars). Then subtract, since the question asks for the difference. If you chose D, you added instead. C and A are partial answers to the problem.

2. **EXPLANATION: The correct answer is 1.93.**

   Start by plugging 100 into the formula you have been given: $(0.0041(100) + 0.676)^8 - 0.000019 (100 + 16) + 0.001316 = x$. Working carefully on your calculator, simplify. The next step looks like this: $(0.41 + 0.676)^8 - 0.000019 (116) + 0.001316 = x$. Then, $1.086^8 - 0.000888 = 1.933$, which can be gridded in as 1.93.

3. **EXPLANATION: The correct answer is 0.06.**

   Each year, the population will grow by 6%, meaning that it will have the whole population from the previous year (represented by the 1) plus an additional 6%, the decimal 0.06. If you thought the answer was 0.6, you should review how to convert percents to decimals. If you put 0.94, you're decreasing the population, not growing it!

4. **EXPLANATION: The correct answer is 6,300.**

   Plugging the numbers into the formula the problem gives: $p = 5,000 (1 + 0.6)^4$. Don't forget to follow the order of operations and to round to the nearest hundred as the problem instructs.

5. **EXPLANATION: Choice C is correct.**

   This is a great problem to work backwards from the answer choices on. Start with 3. If 3 floors were built during the third year, then 1.5 were built during the second year, which doesn't make sense: in real-world problems like this, you often can't have a fractional value. Now try B. If 6 were built in year three, then 3 were built in year two, and 1.5 in the first year: we have the same problem. Working from choice C, 12 built in the third year would mean 6 were built in the second year and 3 in the first year. Work forward from 12 to get 24 in the fourth year, 48 in the fifth year, and 96 in the sixth year. Adding them together gives you 189, the required number of floors, so this is the correct answer. Don't be afraid to work from the answer choices rather than attempting to write a formula for every problem!

6. **EXPLANATION: Choice A is correct.**

   The deposit needs to grow each year, so the number in the parentheses needs to be 1 (representing 100% of the value of the account) plus the percentage by which it grows (0.0375 in this case). A mistake with the decimal point or adding the 1 would have led to selecting a different choice.

7. **EXPLANATION: The correct answer is 2,390.**

   Do the problem in two parts, as shown:

Super Saver CD		Premier CD
$(1,000)(1.035)^5$	+	$(1,000)(1.0375)^5$
$1,187.686		$1,202.099

   Total = $2,389.7858.

   Round to the nearest dollar as the question requires, and disregard the dollar sign.

8. **EXPLANATION: Choice C is correct.**

   Because the growth happens only every 5 years, the exponent t needs to be divided by 5, which leaves only choices A and C. The number in the parentheses needs to be 1.02 to account for 2% growth over the original total.

9. **EXPLANATION: Choice A is correct.**

An exponential function grows or shrinks by some multiple of the current amount, not a constant (which is how a linear function grows). Choices B, C, and D all involve the lake's depth changing by some form of constant.

10. **EXPLANATION: Choice B is correct.**

The line needs to curve downward, which gets rid of choice C. It is not a linear relationship, since it is changing by percent: eliminate D. The drop needs to begin more sharply, not begin more gradually and then drop more steeply as in A. You can graph $y = 100(0.96)\char94 x$ for comparison.

11. **EXPLANATION: Choice A is correct.**

Each year, the number of shares is decreasing, but if you put the 0.024 in the parentheses, you'll be decreasing by a huge amount: over 97% each year. Instead, the number in the parentheses needs to represent the amount that will remain each year, as in choice A. If you're ever not sure on a problem like this, it may help to model the problem on your calculator—seeing the huge and quick drop in the number of shares when you try out choice B will make it immediately clear that it is unlikely to be the answer.

12. **EXPLANATION: Choice C is correct.**

This one can be graphed using points. For example, if $x = 0$, $g(x) = 0$, the graph must go through the origin. That should be enough to do the process of elimination down to C, but don't be afraid to check another point (such as 1, 1) to be sure. You can also try graphing the equation on your calculator if a problem like this one is in a calculator section on the test.

13. **EXPLANATION: Choice B is correct.**

The number is getting smaller, so eliminate the choices that mention growth. Since the number is getting smaller by a different amount each time, rather than by a constant amount, the function is not linear, as in choice A.

14. **EXPLANATION: Choice C is correct.**

Following the format of the equation means choices A and B are eliminated: they do not match. Choice D multiplies the number of months per year, 12, by the number of months, m, rather than dividing. Dividing is necessary to take into account that the amount sold in a month will be less (specifically, less by a factor of 12) than the number of gallons sold in a year.

15. **EXPLANATION: Choice D is correct.**

6 is the initial amount, which needs to grow by 4.2% each year over the previous year. As we've seen in previous questions, the 1, representing 100% of the previous total, needs to be present along with the 0.042 decimal to represent the growth. Choice D is the one that meets these requirements.

CHAPTER 18

DESCRIPTIVE STATISTICS

You will calculate the **mean, median,** and **mode** of a set of numbers. However, the SAT questions will ask you to do more. You may have to find the missing number in a set given the mean, median, or mode, or you may be asked to use the mean, median, or mode to solve another problem. You will also be asked to compute/compare the **range** and **standard deviation** of data sets. The data sets could be given as a list of data, organized in a table, or presented in a graph.

Begin with the mathematics review and then complete and correct the practice problems. There are 2 Solved SAT Problems and 11 Practice SAT Questions with answer explanations.

The **arithmetic mean (average), median,** and **mode** are used to describe a set of data when each item in the set is a number.

> The **arithmetic mean (average)** is the sum of the items divided by the number of items.

> The **median** is the middle number when the list is placed in order, if there is an odd number of items. If there is an even number of items the median is the mean of the middle two numbers.

> The **mode(s)** is (are) the item(s) that occur(s) most frequently. If every item appears the same number of times, then there is no mode in the set.

> The **range** is the distance between the largest value in a data set and the smallest value in a data set.

> The **standard deviation** is best thought of as the average distance of the values in the data set from the mean.

Example 1.

Find the mean, median, mode, and range for each data set.

> a. $\{5, 9, 4, 3, 5, 10, 6\}$
>
> *Mean:*
> $(5 + 9 + 4 + 3 + 5 + 10 + 6) \div 7 = 6$
> 6 is the mean.
>
> *Median:*
> First rewrite the items from least to greatest
> $\{3, 4, 5, \underline{5}, 6, 9, 10\}$
> 5 is the median.
>
> *Mode:*
> The mode is 5; it appears most often.
>
> *Range:*
> Largest value = 10 and Smallest value = 3
> Range = 10 − 3 = 7.

b. {63, 67, 54, 68, 76, 54, 87, 63}

Mean:

$(63 + 67 + 54 + 68 + 76 + 54 + 87 + 63) \div 8 = 66.5$

66.5 is the mean.

Median:

First rewrite the items from least to greatest

{54, 54, 63, 63, 67, 68, 76, 87}.

Because there are an even number of items, find the mean of the middle two numbers:

$$\frac{63 + 67}{2} = 65$$

65 is the median.

Mode:

In this problem there are two modes: 54 and 63. This is because both 54 and 63 occur most frequently.

Range:

Largest value = 87 and Smallest value = 54

Range = 87 − 54 = 33

It is possible for two sets of data to have the same range but different standard deviations.

Example 2.

The tables below give the distribution for two data sets. Find the mean, median, mode, and range for each data set. Then compare the standard deviations of the data set.

Data Set A

Value	Frequency
1	5
2	5
3	7
4	5
5	5

Data Set B

Value	Frequency
1	2
2	3
3	17
4	3
5	2

Mean A:

$$\frac{5(1) + 5(2) + 7(3) + 5(4) + 5(5)}{27} = 3$$

Median A: 3 is the middle number.

Mode A: 3 is the most frequent number.

Range A: 5 − 1 = 4.

Mean B:

$$\frac{2(1) + 3(2) + 17(3) + 3(4) + 2(5)}{27} = 3$$

Median B: 3 is the middle number.

Mode B: 3 is the most frequent number.

Range A: 5 − 1 = 4.

Most of the values in Data Set B are at the mean, while the values in Data Set A are more spread out. It follows that the standard deviation for Data Set B is smaller than the standard deviation for Data Set A.

Example 3.

The dot plots below show the stop distance, in feet, for 10 different tires under dry and wet conditions. Which data set has a smaller standard deviation?

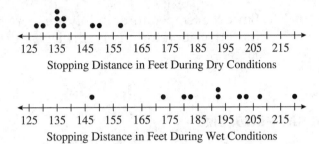

Stopping Distance in Feet During Dry Conditions

Stopping Distance in Feet During Wet Conditions

Stopping distance during dry conditions has a smaller standard deviation. This is because the graph of stopping distance during dry conditions are grouped closer together than the graph of stopping distance during wet conditions.

Practice Questions

1. On three of the first four geometry tests, Jim earned the following scores: 84, 92, and 88. If Jim's average for all four tests is 87.5, what score did Jim earn on his fourth test?
2. There are 24 students in period 1 gym class and 20 students in a period 2 gym class. The students ran an obstacle course during gym class. The average time in period 1 was 165 seconds, and the average time in period 2 was 180 seconds. What is the average time for the two combined gym classes?
3. The histograms below show the average temperature, in degrees Fahrenheit (°F), from 35 different countries during the months of January and July. In which month is the standard deviation of temperatures larger?

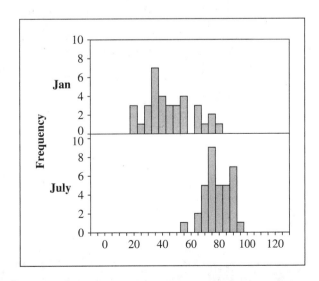

Practice Answers

1. $\dfrac{84 + 92 + 88 + x}{4} = 87.5 \;\Rightarrow\; 84 + 92 + 88 + x = 350$

   $264 + x = 350 \;\Rightarrow\; x = 86.$

2. $\dfrac{24 \cdot 165 + 20 \cdot 180}{44} = 171.\overline{81}$ seconds.

3. Since the temperatures in July are more tightly clustered together, the standard deviation for July is smaller, so the standard deviation for January is larger.

SOLVED SAT PROBLEMS

1.

Derek	$4.29	$6.99	$5.29	$1.99	x
Ann	$6.29	$2.49	$3.99	$0.99	$3.19

Derek and Ann each purchased 5 items from the grocery store, and the price of each item is shown in the table. The mean amount of money that Derek spent on her 5 items is $0.50 more than the mean amount the Ann spent on his 5 items. What is the value of x?

EXPLANATION: The correct answer is 0.89.

First compute the mean amount Brendan spent.

$$\dfrac{6.29 + 2.49 + 3.99 + 0.99 + 3.19}{5} = 3.39$$

Therefore, Derek's mean spent must be $3.39 + 0.5 = 3.89$. Since Derek also bought 5 items, her total spent must be $5(3.89) = 19.45$.

Add Derek's purchases and set equal to 19.45:

$$4.29 + 6.99 + 5.29 + 1.99 + x = 19.45$$

Simplify expression: $18.56 + x = 19.45$

Subtract 18.56 from both sides: $x = 0.89$.

2. The Men's Giant Slalom skiing event consists of two runs whose times are added together for a final time. The data shown in the graph below gives the final giant slalom times, in seconds, from the 2014 Winter Olympics at Sochi. A total of 72 final times were reported. Which of the following correctly describes what we can tell about whether the mean or the median is greater based on the data displayed in the graph?

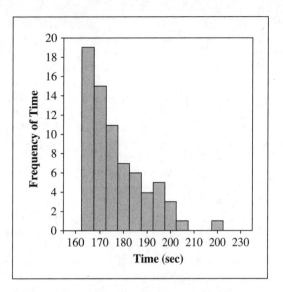

A. The mean and median will be spread apart because of the cluster of values on the right half of the graph.

B. The mean and median will be relatively close to each other because of the cluster of values on the right half of the graph.

C. The mean and median will be relatively close to each other because of the cluster of values on the left half of the graph.

D. The median and mean will be spread apart because of the cluster of values on the left half of the graph.

EXPLANATION: Choice C is correct.

Note that with grouped data like this you do not know the individual scores, and so you cannot find the actual mean, median and mode. There are 72 values, so the median is the average of the 36th and 37th values. The first two bars have a total of 19 + 15 = 34 values. Therefore, the median lies within the 3rd bar. The mean will be shifted left as well, and so the median and the mean will be relatively close to each other because of the cluster of bars to the left side of the graph.

DESCRIPTIVE STATISTICS
PRACTICE SAT QUESTIONS

ANSWER SHEET

Choose the correct answer.

1. Ⓐ Ⓑ Ⓒ Ⓓ   11. Ⓐ Ⓑ Ⓒ Ⓓ
2. Ⓐ Ⓑ Ⓒ Ⓓ
3. Ⓐ Ⓑ Ⓒ Ⓓ
4. Ⓐ Ⓑ Ⓒ Ⓓ
5. Ⓐ Ⓑ Ⓒ Ⓓ
6. Ⓐ Ⓑ Ⓒ Ⓓ
7. Ⓐ Ⓑ Ⓒ Ⓓ
8. Ⓐ Ⓑ Ⓒ Ⓓ
9. Ⓐ Ⓑ Ⓒ Ⓓ
10. Ⓐ Ⓑ Ⓒ Ⓓ

PRACTICE SAT QUESTIONS

1. The average of x and y is 7, and $z = 3x + 2$. What is the average of y and z?

   A. $2x + 8$
   B. $2x - 16$
   C. $x - 8$
   D. $x + 8$

2. $\dfrac{a + b + c + d}{4} = 12$. If the average of a, b, c, d, and e is 14, what is the value of e?

   A. 14
   B. 16
   C. 18
   D. 22

3. There were 10 softball teams in a tournament. The mean number of runs for all the teams is 6.4. Only 8 of the teams made it to the playoffs. The average number of runs for the teams that made it to the playoffs was 7.1 rounded to the nearest tenth. What was the combined number of runs for the two teams that did not make it to the tournament?

   A. 5
   B. 6
   C. 7
   D. 8

4. The tables below show the points scored by two different basketball players, Jay and Ed. Ed's average (arithmetic mean) for 4 games is 2 less than Jay's average for 5 games.

Jay's points	Ed's points
20	15
16	21
25	17
15	
14	

   How many points did Ed score during the fourth game?

   A. 12
   B. 11
   C. 10
   D. 9

5. A total of 25 golfers participated in a one-day tournament. The frequency table below shows the distribution of scores.

GOLF	SCORES
74	1
73	5
71	1
70	2
69	4
68	2
67	4
66	4
65	2

   What is the sum of the mean, median, and mode of the scores rounded to the nearest whole number?

   A. 203
   B. 208
   C. 211
   D. None of the above

6. Trains A and B travel between the same two stations. For 20 trips, conductors on each train keep track of the number of minutes each train is late to the nearest 5 minutes. The results of these efforts are shown on the dot plots below. Which of the following is the best description of the comparison of the patterns shown in the plots?

   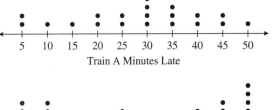
   Train A Minutes Late

   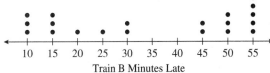
   Train B Minutes Late

   A. The standard deviations and ranges are different, but the medians are the same.
   B. The ranges are the same, but the medians and standard deviations are different.
   C. The standard deviations are the same, but the medians and ranges are different.
   D. The medians and ranges are the same, but the standard deviations are different.

7. Each morning for 30 days during ski season, a member of the ski patrol measures snow depth to the nearest foot. The readings are recorded in the table below.

Ski Slope A	
Snow Depth	Frequency
5	5
4	5
3	10
2	6
1	4

Ski Slope B	
Snow Depth	Frequency
5	8
4	9
3	6
2	4
1	3

Which of the following is true about the data recorded in these tables?

A. The standard deviations in each table are essentially the same.
B. The standard deviation for slope A is larger.
C. The standard deviation for slope B is larger.
D. The sample size in each table is too small to answer any questions.

8. The histogram below shows attendance in thousands at 10 different performances.

Thousands of Fans at 10 Different Performances

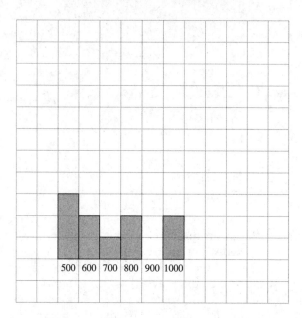

What is the mode and what is the median of the data from the histogram above?

A. Mode 6,000; median 5,000
B. Mode 5,000; median 6,500
C. Mode 6,500; median 5,000
D. Mode 10,000; median 6,500

9. In a review of the cost of a gallon of gasoline from all service stations in a county, it was found that the mode of the price of a gallon of gas was much lower than the average for the price of a gallon of gas. Which of the following would explain the reason for this difference?

A. The average price of a gallon of gas was based on all the gas sold, but the mode was just based on some of the prices.
B. The mode shows a cluster of prices for the lowest prices, while the mean shows a result for all the prices.
C. Mode and mean are both measures of central tendency, and it just happens that in this case the mode is lower.
D. The mean will always be more than the mode because the mean includes higher prices.

10. The real estate appraiser completes visits to all the residences in Elk Village and Deer Village, which make up the entire town of Grand Village. The number of bedrooms in each residence is noted below. Studio residences are recorded with no bedrooms.

Number of Bedrooms in
Grand Village Residences

Number of Bedrooms	Elk Village	Deer Village
0 (Studio)	80	60
1	200	80
2	60	180
3	50	20
4	5	0

Which of the following is the median of the number of bedrooms in Grand Village?

A. 1
B. 2
C. 3
D. 4

11. There are two nearby but larger towns: Moose Village and Antelope Village. Proportionally, the data from Elk Village is a good match for Moose Village and the data from Deer Village is a good match for Antelope Village, as is the data for Moose Village and Antelope Village. Using the information about Grand Village from the table given, which is the best conclusion about the number of residences with four bedrooms in Moose Village and Antelope Village?

A. There will be more four-bedroom residences in Moose Village than in Antelope Village.
B. There will be more four-bedroom residences in Antelope Village than in Moose Village.
C. There will be about the same number of four-bedroom residences in Antelope Village and Moose Village.
D. There will be more four-bedroom residences in Deer Village versus Moose Village.

▉ EXPLAINED ANSWERS

1. **EXPLANATION: Choice D is correct.**

   Write an equation for the average of x and y.

   $$\frac{x+y}{2} = 7 \quad \text{Solve for } y \quad \Rightarrow \quad x+y = 14 \quad \Rightarrow \quad y = 14 - x.$$

   The problem states $z = 3x + 2$.

   Write an equation for the average of y and z.

   $$\frac{y+z}{2} = \frac{(14-x)+(3x+2)}{2} = \frac{16+2x}{2}$$
   $$\frac{2(8+x)}{2} = 8 + x = x + 8.$$

2. **EXPLANATION: Choice D is correct.**

   The average of a, b, c, and d is 12, so the sum of a, b, c, and d is four times the average, or 48. The sum of a, b, c, d, and e is 5 times the average, or 70.

   Subtract to find e.

   $70 - 48 = 22$, so $e = 22$.

3. **EXPLANATION: Choice C is correct.**

   Multiply 6.4 by 10 to figure out there were 64 runs scored in total by the 10 teams. Then multiply 7.1 by 8 and round 56.8 up to 57 to find the number of runs scored by the 8 tournament teams. Then subtract: $64 - 57 = 7$.

4. **EXPLANATION: Choice B is correct.**

   Write an equation that shows the relationship between Ed's points and Jay's points. The variable x represents the missing score from Ed's points. Solve the equation.

   $$\frac{15+21+17+x}{4} = \frac{20+16+25+15+14}{5} - 2$$
   $$\frac{53+x}{4} = 18 - 2 \quad \Rightarrow \quad \frac{53+x}{4} = 16$$
   $$53 + x = 64 \quad \Rightarrow \quad x = 11$$

   Ed scored 11 points in the fourth game.

5. **EXPLANATION: Choice C is correct.**

   Start by finding the mean: there is 1 score of 75, there are 5 scores of 73, 1 score of 71, and so on. Divide your total, 1724, by 25 to get 69 as the mean. The mode is score with the highest frequency, so that's 73, which occurred five times. Finally, find the median: there are 25 scores—you can write them all out or use the frequency table to count down until you reach the 13th, or middle score. That score is 69. Now add the three numbers together to get 211.

6. **EXPLANATION: Choice B is correct.**

   The ranges of the two dot plots are both 45, but the medians are different: Train A is 30 and Train B is 37.5. That eliminates choices A and C, so you can take a good guess even without knowing how to calculate standard deviation. Standard deviation is essentially about where the data is clustered: because these two plots are arranged so differently, the standard deviations are clearly different.

7. **EXPLANATION: Choice C is correct.**

The mean of both slopes is a bit over 3, but the values of slope A are clustered more tightly around that mean, with many of the values being 3 and many more at 2 or 4. The values of slope B are spread out further, with a lot being clustered near the top of the range: thus, the standard deviation of slope B is larger. Remember that the SAT won't actually ask you to calculate standard deviation, so being able to find the mean and see how the scores are clustered is a strategy that will always work.

8. **EXPLANATION: Choice B is correct.**

The highest bar on the graph is at the 5,000 value, so that's the mode. The median is between the fifth and sixth performance if you're estimating, which gives you 6,500, as in choice B. Note that, as here, you may not need to calculate both values fully to find the credited response.

9. **EXPLANATION: Choice B is correct.**

The definition of mode is that it's the most common data point, whereas mean, or average, takes into account all of the data points. There is no evidence for choice A, and choice C is unlikely to ever be a correct answer, though it is a distractor choice you may see: remember that the SAT isn't likely to have "coincidence" rather than "math" be the right answer! Although choice D may seem possible at first glance, it doesn't make sense given the correct definition of mode. The mode is the price that occurs most often, and it is likely in this case that the mode represents a cluster of gas prices that are lower than the mean.

10. **EXPLANATION: Choice A is correct.**

Find the combined results of Elk Village and Deer Village, as seen here:

Number of Bedrooms	Elk Village		Deer Village	
0 (Studio)	80	+	60	= 140
1	200	+	80	= 280
2	60	+	180	= 240
3	50	+	20	= 70
4	5	+	0	= 5
			TOTAL	735

There are 735 residences in Grand Village. The total is an odd number, so find the position of the median by dividing $\frac{(735 + 1)}{2} = \frac{736}{2} = 368$. Use the frequency table and cross out values or add down to see that the median is 1.

11. **EXPLANATION: Choice A is correct.**

The number of four-bedroom residences in Elk and Deer Villages is significantly different, as there are more four-bedroom residences in Moose Village compared to Antelope Village. Therefore, because there are 5 times the number of apartments between Elk village and Deer Village, we would expect this proportion to be the case for Moose Village versus Antelope Village—that is, 5 times the number of residences. Choice B is incorrect, because it has the opposite villages. Choice C is incorrect, because there is a clear difference in the number of residences between the two. Finally, choice D is incorrect, because it is comparing the incorrect villages and it is still incorrect as it is essentially choice B. Note that you don't need to have actual numerical values to do this problem, which is often true of descriptive statistics on the SAT!

CHAPTER 19

PROBABILITY

This chapter reviews probability of events, the meaning of and/or, and particularly focuses on two-way table SAT problems. Begin with the mathematics review and then complete and correct the practice problems. There are 2 Solved SAT Problems and 10 Practice SAT Questions with answer explanations.

$$\text{The probability of an event} = P(E) = \frac{\text{Number of ways the event can occur}}{\text{Total number of possible outcomes}}$$

Example 1.

There are 60 red marbles, 40 green marbles, and 50 blue marbles in a jar. A marble is picked at random. What is the probability that the marble is red?

$$P(\text{Red}) = \frac{60}{150} = \frac{6}{15} = \frac{2}{5} = 0.4.$$

What is the probability that the marble is blue and green?

$P(\text{Blue and Green}) = 0$. None of the marbles are both blue and green.

What is $P(\text{Blue or Green}) = \dfrac{60 + 40}{150} = \dfrac{100}{150} = \dfrac{2}{3} = 0.66$

What is $P(\text{Not Green}) = \dfrac{150 - 40}{150} = \dfrac{110}{150} = \dfrac{11}{15} = 0.73$

Use this table for Examples 2–6.

The table below shows the distribution of political affiliation and gender from a survey of 202 individuals. Use the results to answer the questions below.

Gender	High School Student Population			TOTAL
	Sophomores	Juniors	Seniors	TOTAL
Female	48	33	16	97
Male	36	24	45	105
TOTAL	84	78	40	202

Example 2.

What is the probability that a student chosen at random from the survey is female?

There is a total of 202 students in the survey, and 97 of them are female.

$$P(Female) = \frac{97}{202}.$$

Example 3.

What proportion of students in the survey are not Sophomores?

There is a total of 202 students in the survey, and 84 of them are Sophomores. Therefore, 202 − 84 = 118 are not Sophomores.

$$P(Not\ Sophomore) = \frac{118}{202}.$$

Example 4.

What is the probability that a student chosen at random from the survey is male and a Senior?

There are a total of 202 students in the survey, and 45 are male and Senior.

$$P(Male\ and\ Senior) = \frac{45}{202}.$$

Example 5.

What proportion of Juniors in the survey are female?

There are a total of 78 Juniors in the survey, and 33 of them are female.

$$P(Female\ and\ Junior) = \frac{33}{78}.$$

Example 6.

What is the probability that a student chosen at random from the survey is male or Junior?

There are a total of 202 students in the survey. 105 are male, 78 are Juniors, and 24 are both male and Juniors. To make sure not to count people twice, we must add the Male group to the Junior group and subtract those that are both.

$$P(Male\ or\ Junior) = \frac{105 + 78 - 24}{202} = \frac{159}{202}.$$

Practice Questions

The data in the table below were produced by a researcher studying the relationship of a man's age and his blood pressure from a sample of 474 men. Use the results to answer the questions below.

	Age			
Blood Pressure	**Under 30**	**30–49**	**50 or Over**	**TOTAL**
Low	23	51	73	147
Normal	27	37	31	95
High	48	91	93	232
TOTAL	98	179	197	474

1. What proportion of men from the study are 50 or over with normal blood pressure?

2. Of the men in the study with high blood pressure, what fraction of them are under 30?
3. What is the probability that a man chosen from the study is either between 30 and 49 years of age or has low blood pressure?
4. If a man is chosen at random from those in the study is under 30, what is the probability that he has normal blood pressure?

Practice Answers

1. $P(50 \text{ or Over with Normal Blood Pressure}) = \dfrac{31}{474}$.

2. $P(\text{Under 30 from those with High Blood Pressure}) = \dfrac{48}{232}$.

3. $P(30\text{–}49 \text{ or Low Blood Pressure}) = \dfrac{275}{474}$.

4. $P(\text{Normal Blood Pressure from those Under 30}) = \dfrac{27}{98}$.

SOLVED SAT PROBLEMS

School Affiliation	Car Origin		TOTAL
	Foreign	**Domestic**	
Staff	88	107	195
Student	59	105	164
TOTAL	147	212	359

1. What fraction of the cars are student cars or domestic cars?

   A. $\dfrac{105}{359}$

   B. $\dfrac{271}{359}$

   C. $\dfrac{212}{359}$

   D. $\dfrac{105}{212}$

   EXPLANATION: Choice B is correct.

   There are 164 student cars and 212 domestic cars. Add $164 + 212 = 376$.

   There are 105 cars that are both student and domestic. Subtract that number of cars. $376 - 105 = 271$.

   Now divide by the total number of cars.

   $P(\text{Student or Domestic}) = \dfrac{271}{359}$.

Championship Game Viewing	Gender		TOTAL
	Female	Male	
Game	200	279	479
Commercials	156	81	237
Don't Watch	160	132	292
TOTAL	516	492	1,008

2. 1,008 people over the age of 18 were asked in a survey if they planned to watch the championship game, and if so, if they were more interested in the game or the commercials. The results are shown in the table above. If a person is chosen at random, what is the probability the person will be either a male who does not watch the championship game or a female who watches the championship game, rather than the commercials?

A. $\dfrac{163}{252}$

B. $\dfrac{265}{336}$

C. $\dfrac{83}{252}$

D. $\dfrac{257}{336}$

EXPLANATION: Choice C is correct.

There are a total of 132 males who responded they would not watch the championship game and 200 females who responded who watch for the game and not the commercials. That is a total of 332 out of the 1,008 in the survey.

$$P(\text{Male Does Not Watch or a Female Watches for Game}) = \frac{332}{1,008} = \frac{83}{252}.$$

PROBABILITY
PRACTICE SAT QUESTIONS

▨ ANSWER SHEET

Choose the correct answer.
If no choices are given, grid the answers in the section at the bottom of the page.

1. Ⓐ Ⓑ Ⓒ Ⓓ
2. Ⓐ Ⓑ Ⓒ Ⓓ
3. Ⓐ Ⓑ Ⓒ Ⓓ
4. Ⓐ Ⓑ Ⓒ Ⓓ
5. Ⓐ Ⓑ Ⓒ Ⓓ
6. **GRID**
7. Ⓐ Ⓑ Ⓒ Ⓓ
8. Ⓐ Ⓑ Ⓒ Ⓓ
9. Ⓐ Ⓑ Ⓒ Ⓓ
10. Ⓐ Ⓑ Ⓒ Ⓓ

Use the answer spaces in the grid below if the question requires a grid-in response.

Student-Produced Responses ONLY ANSWERS ENTERED IN THE CIRCLES IN EACH GRID WILL BE SCORED. YOU WILL NOT RECEIVE CREDIT FOR ANYTHING WRITTEN IN THE BOXES ABOVE THE CIRCLES.

6.

	/	/	/
.	.	.	.
	0	0	0
1	1	1	1
2	2	2	2
3	3	3	3
4	4	4	4
5	5	5	5
6	6	6	6
7	7	7	7
8	8	8	8
9	9	9	9

PRACTICE SAT QUESTIONS

1. 60 blue marbles and 40 red marbles are in a jar. How many red marbles must be removed from the jar so that the probability of choosing a blue marble from the jar is $\frac{3}{4}$?

   A. 5
   B. 10
   C. 15
   D. 20

2. The figure below consists of two congruent semicircles on either end of a square. What is the probability, rounded to the nearest hundredth, that a point in the region chosen at random is not in the shaded area?

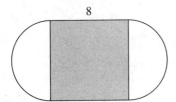

   A. 0.56
   B. 0.48
   C. 0.46
   D. 0.44

3. Three coins are tossed at the same time. What is the probability that exactly two heads are face up?

   A. $\frac{1}{8}$

   B. $\frac{1}{4}$

   C. $\frac{3}{8}$

   D. $\frac{1}{2}$

4. A jar contains 10 blue, 8 green, and 6 red marbles. Every time a marble is removed from the jar, it is not replaced. What is the probability, to the nearest hundredth, that the second marble chosen is green if the first marble chosen is green?

   A. 0.28
   B. 0.29
   C. 0.30
   D. 0.31

5. Blaire, Chad, Ellen, and Jordan are randomly arranged in four seats at the front row of the classroom. What is the probability that Chad and Blaire are sitting next to each other?

   A. 0.15
   B. 0.25
   C. 0.4
   D. 0.5

6. A dart is thrown randomly at the target below. The radius of the innermost circle is 2, and the radius of each circle doubles as the circles get bigger. What is the probability the dart will hit the shaded region, to the nearest hundredth of a percentage?

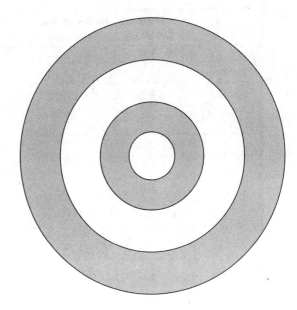

7. $X = \{-4, -2, 1, 3\}$ $Y = \{-1, 4, 5\}$. If x is a number from set X, and y is a number from set Y, the probability that $x + y$ is positive is closest to

   A. 0.5
   B. 0.6
   C. 0.7
   D. 0.8

8.

Answer	Percent
Strawberry	34.2%
Vanilla	15.1%
Other	40.2%
I don't eat ice cream.	10.5%

The table above shows the data collected when students at a local school were questioned about their favorite ice cream flavor. Based on this information, what is the approximate probability that a student answered Vanilla, given that he or she did not say "I don't eat ice cream"?

A. 0.0151
B. 0.105
C. 0.151
D. 0.169

9.

	Additional Self-study	No Additional Self-study	Total
Test prep class online	35	12	47
Test prep class in person	40	15	55
Total	75	27	102

A college surveyed 102 members of its freshman class. The table shows whether the students reported taking a test prep class in person or online for the SAT and whether or not they engaged in additional self-study on their own. Based on this information, what is the probability that a randomly selected freshman took an in-person test prep class?

A. $\frac{55}{102}$

B. $\frac{40}{102}$

C. $\frac{75}{102}$

D. $\frac{15}{55}$

10.

	Weight Loss	No Weight Loss	Total
Nutritionist	125	75	200
Personal trainer	80	120	200
Total	205	195	400

A research study examined whether people lost weight using various methods. 400 adults in a random sample were assigned either a nutritionist or a personal trainer to help them lose weight. After six weeks, the study participants reported whether or not they lost weight. What fraction of the students who reported weight loss were assigned a nutritionist?

A. $\frac{16}{41}$

B. $\frac{25}{41}$

C. $\frac{24}{39}$

D. $\frac{41}{80}$

1. **EXPLANATION: Choice D is correct.**

   If 20 red marbles are removed, there are 60 blue marbles and 20 red marbles remaining.

   $P(\text{Blue}) = \dfrac{60}{80} = \dfrac{6}{8} = \dfrac{3}{4}$.

2. **EXPLANATION: Choice D is correct.**

   The two semicircles, which make up a single circle, have the same diameter as one side of the square, $d = 8$ and $r = 4$. That means the area of the entire figure is $8^2 + \pi(4)^2 = 64 + 16\,\pi$.

   The area of the nonshaded region, which is the two semicircles, is $\pi(4)^2 = 16\,\pi$.

   The probability of choosing a point not in the shaded region is the probability of not choosing a point in the shaded region. $\dfrac{16\,\pi}{64 + 16\,\pi} \approx 0.44$.

3. **EXPLANATION: Choice C is correct.**

   First, list all the possible outcomes when three coins are tossed in the air. When a coin is tossed, there are two outcomes, heads or tails. When three coins are tossed, there are $2 \times 2 \times 2 = 8$ possible outcomes $\{HHH, HHT, HTH, THH, HTT, THT, TTH, TTT\}$.

   Three of these outcomes have exactly two heads.

   Therefore, $P(\text{Exactly Two Heads}) = \dfrac{3}{8}$.

4. **EXPLANATION: Choice C is correct.**

   Because the first marble drawn is green that means there are a total of 23 marbles remaining. Seven of these are green.

   $P(\text{Second Green}) = \dfrac{7}{23} \approx 0.30$.

5. **EXPLANATION: Choice D is correct.**

   There are a total of $4 \times 3 \times 2 \times 1 = 24$ ways to arrange four people.

   Let B = Blaire, C = Chad, E = Ellen, and J = Jordan. The different arrangements where Chad is next to Blaire are: BCEJ, BCJE, CBEJ, CBJE, JBCE, EBCJ, JCBE, ECBJ, EJBC, JEBC, EJCB, and JECB. There are 12 ways for Chad to be seated next to Blaire.

   $P(\text{Chad next to Blaire}) = \dfrac{12}{24} = \dfrac{1}{2} = 0.5$.

6. **EXPLANATION: The correct answer is 0.80.**

The outer shaded region has an area of $\pi(16)^2 - \pi(8)^2 = 256\pi - 64\pi = 192\pi$.

The inner shaded region has an area of $\pi(4)^2 - \pi(2)^2 = 16\pi - 4\pi = 12\pi$.

The total shaded area is $192\pi + 12\pi = 204\pi$.

The area of the entire region is $\pi(16)^2 = 256\pi$.

$$P(\text{shaded area}) = \frac{204\,\pi}{256\,\pi} = 0.80.$$

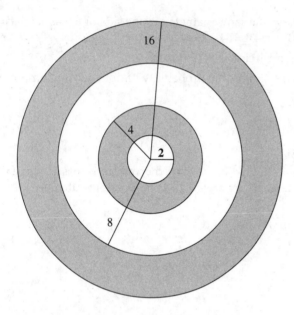

7. **EXPLANATION: Choice C is correct.**

Split the problem into four cases, one for each possible value of x. Then add each possible value of y $(-1, 4, 5)$.

$x = -4$	$x = -2$
$x + y = -4 + -1 = -5$	$x + y = -2 + -1 = -3$
$x + y = -4 + 4 = 0$	$x + y = -2 + 4 = 2$
$x + y = -4 + 5 = 1$	$x + y = -2 + 5 = 3$
$x = 1$	$x = 3$
$x + y = 1 + -1 = 0$	$x + y = 3 + -1 = 2$
$x + y = 1 + 4 = 5$	$x + y = 3 + 4 = 7$
$x + y = 1 + 5 = 6$	$x + y = 3 + 5 = 8$

There are 12 answers to $(x + y)$. Eight of these are greater than 0.

The probability that the sums of x and y are positive (greater than zero) is $\frac{8}{12} = 0.66\ldots$, which is closest to 0.7.

8. **EXPLANATION: Choice D is correct.**

The problem specifies that we aren't counting the people who don't eat ice cream, so the fraction we want is 15.1 (the vanilla group) over a total that doesn't include those people (89.5). $\frac{15.1}{89.5} = 0.168$, or 1,687, which rounds to 0.169. If you put choice C, you misread the question and gave the percentage of vanilla eaters, and if you selected A, you made the same mistake with an additional error in converting percent to decimal.

9. **EXPLANATION: Choice A is correct.**

The total number of students who took a test prep class in person is 55, out of a total of 102. The other answer choices result from using the wrong data from the table. If you thought 55 should be on the bottom of the fraction, as in choice D, you were looking at the population of students who took a test prep class in person as your total population, rather than as fraction of the total, which is what the question specified.

10. **EXPLANATION: Choice B is correct.**

The fraction in question is $\frac{125}{205}$, and it needs to be reduced. If you used the wrong total or wrong numerator for the fraction, you might have gotten one of the answer choices: these are typical SAT distractors in that they're not random but can be avoided by a close reading of the question.

CHAPTER 20

STATISTICAL GRAPHS, SCATTER PLOTS, AND LINES OF BEST FIT

SAT problems on **scatter plots** and other statistical graphs often ask you to approximate percent change in value, make predictions using the **line of best fit**, interpret the slope of the best-fitting line, and approximate the difference between predicted and actual values. You will be asked questions about appropriate conclusions that can be drawn from a survey based on a random sample. Begin with the mathematics review and then complete and correct the practice problems. There are 2 Solved SAT Problems and 10 Practice SAT Questions with answer explanations.

Bar graph: a bar graph is data represented by rectangles, and can be used to compare data.

Example 1.

The bar graphs below show the land area, in square miles, for six different states. Which of these states has the largest land area?

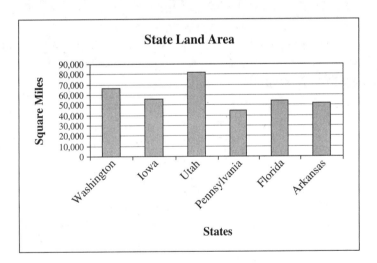

Both bar graphs show the same information. Utah has the longest graph, so it has the largest land area.

Example 2.

In what percent of the months does Portland, Oregon, have fewer rainy days than Portland, Maine?

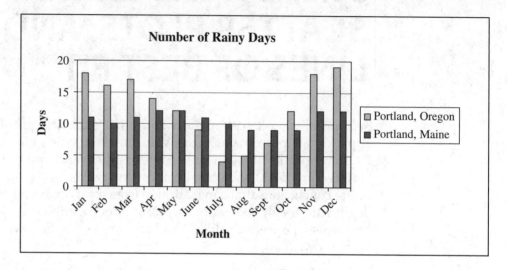

Answer: 33.3

Portland, Oregon, has fewer rainy days in June, July, August, and September. The percent of the months Portland, Oregon, has fewer rainy days than Portland, Maine, is

$$\frac{4}{12} = \frac{1}{3} = 33.3.$$

Line Graph: A line graph shows how data varies, often over time.

Example 3.

The line graph below shows the number of home runs for Barry Bonds from 1993 to 2003. What was his percent decrease in home runs from 2001 to 2002? Divide the decrease in home runs by the number of home runs in 2001.

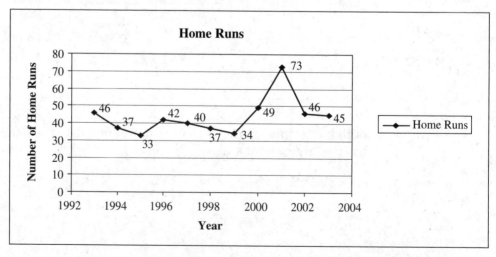

$$\text{Percent Decrease} = \frac{73 - 46}{73} = \frac{27}{73} = 0.37 = 37\%$$

A **scatter plot** shows the relationship between two numeric variables. If the data in general move up and to the right, then the data have a positive correlation, which is sometimes called **direct variation.** If the data move down and to the right, then the data have a negative correlation, meaning one values goes up as the other goes down. This is also known as **inverse variation.**

Example 4.

The scatterplot below shows the maximum speed and maximum height from 14 different roller coasters along with the line of best fit. Approximate and interpret the slope of the best-fitting line to the data.

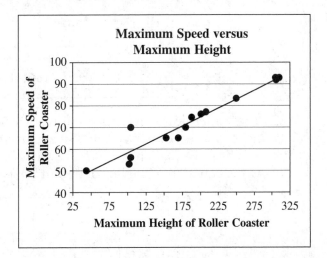

Look for two locations where the point of the best-fitting line can be approximated. There can be several good locations, but right away the line of best fit looks to go through (50, 50) and (75, 54). Therefore, the approximation for the value of the slope of the slope is $\dfrac{54 - 50}{75 - 50} = \dfrac{2}{25} = 0.08$. So, the average max speed of the roller coasters will increase by 0.08 mph for every increase in 1 foot of max height.

Practice Questions

1. The bar graph below shows the distribution of gender based on political affiliations. Rounded to the nearest tenth, what percent of males are Republican?

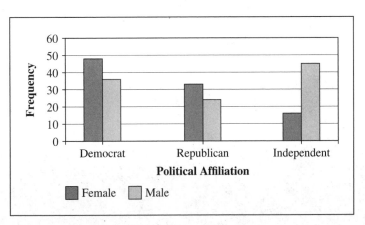

2. The scatterplot below shows the total calories and total fat from 7 different hamburgers. For the hamburger with the highest fat content, approximate the difference between the predicted fat content and the actual fat content.

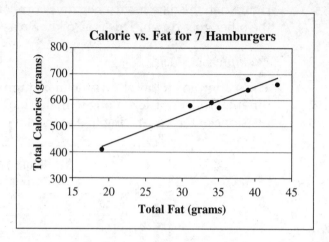

3. The scatterplot below shows the average 30-year fixed mortgage interest rate, in percent, each year from 1985 through 2015. The line of best fit is also shown and has an equation $y = -0.2192z + 10.277$. What is the meaning of the 10.277 in the equation of the line of best fit?

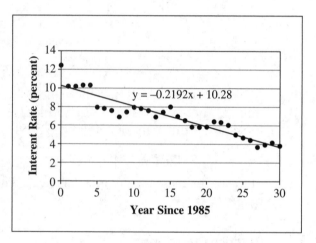

Practice Answers

1. $\dfrac{24}{105} \approx .229 = 22.9\%$.

2. 690 grams – 660 grams = 30 grams.

3. In 1985, the average 30-year mortgage rate was about 10.28%.

SOLVED SAT PROBLEMS

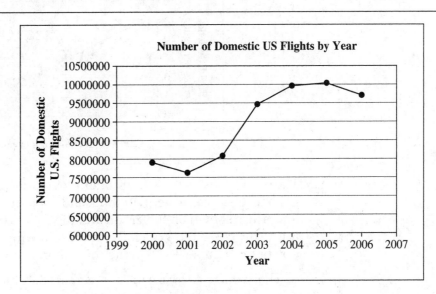

1. According to the line graph above, between what two consecutive years was there the least change in the number of domestic U.S. flights?

   A. 2000–2001

   B. 2002–2003

   C. 2004–2005

   D. 2005–2006

   EXPLANATION: Choice C is correct.

   The problem is asking for the interval where the change, the slope, is the shallowest. Looking at the graph it can be seen that this occurs between 2004 and 2005.

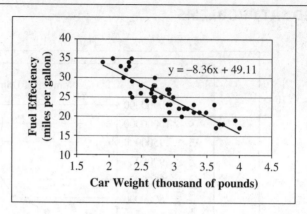

2. The graph above shows the scatter plot of fuel efficiency, in miles per gallon, versus car weight, in thousands of pounds. The best-fitting line is also shown on the graph and the equation of the line of best fit is $y = -8.36x + 49.11$. Which the of the following is the best interpretation of the number -8.36 in the context of the problem?

A. The average decrease in car weight is 8.36 thousand pounds.

B. For every mile per gallon, the average decrease in car weight is 8.36 thousand pounds.

C. The average decrease in car fuel efficiency is 8.36 miles per gallon.

D. For every thousand pounds, the average decrease in car fuel efficiency is 8.36 miles per gallon.

EXPLANATION: Choice D is correct.

-8.36 is the slope of the best-fitting line. The slope for this line is the change in fuel efficiency based on car weight. Looking at the different answer choices, notice that only choice D describes the change in car fuel efficiency, miles per gallon, for every change in car weight, thousands of pounds.

STATISTICAL GRAPHS, SCATTER PLOTS, AND LINES OF BEST FIT PRACTICE SAT QUESTIONS

ANSWER SHEET

Choose the correct answer.
If no choices are given, grid the answers in the section at the bottom of the page.

1. Ⓐ Ⓑ Ⓒ Ⓓ
2. Ⓐ Ⓑ Ⓒ Ⓓ
3. Ⓐ Ⓑ Ⓒ Ⓓ
4. Ⓐ Ⓑ Ⓒ Ⓓ
5. Ⓐ Ⓑ Ⓒ Ⓓ
6. Ⓐ Ⓑ Ⓒ Ⓓ
7. Ⓐ Ⓑ Ⓒ Ⓓ
8. GRID
9. GRID
10. Ⓐ Ⓑ Ⓒ Ⓓ

Use the answer spaces in the grids below if the question requires a grid-in response.

| Student-Produced Responses | ONLY ANSWERS ENTERED IN THE CIRCLES IN EACH GRID WILL BE SCORED. YOU WILL NOT RECEIVE CREDIT FOR ANYTHING WRITTEN IN THE BOXES ABOVE THE CIRCLES. |

8.

9.

■ PRACTICE SAT QUESTIONS

1. According to the graph, between which two months was there the smallest change in the number of new cases of the flu?

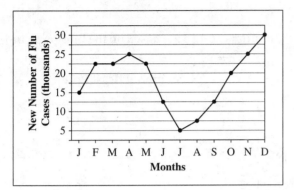

A. February and March
B. March and April
C. July and August
D. August and September

2. The graph below represents the average points per game of a certain basketball player from 1995–2000. What is the percent increase over the 6-year period to the nearest hundredth of a percent?

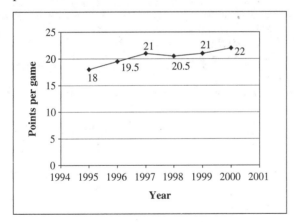

A. 18.18%
B. 20%
C. 22%
D. 22.22%

3. The bar graph below displays the data gathered when individuals were polled about their favorite type of movie. Each person polled chose one movie type.

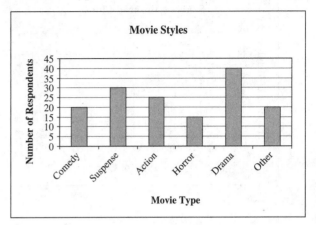

What percent of the respondents chose action or horror?

A. $36\dfrac{2}{3}\%$

B. $33\dfrac{1}{3}\%$

C. $26\dfrac{2}{3}\%$

D. $23\dfrac{1}{3}\%$

4. A total of 250 students participate in five different school-sponsored clubs. The bar graph below shows the number of students who participate in each club. 40% of the members of Club A are in Club B. How many of the members in Club A are not in Club B?

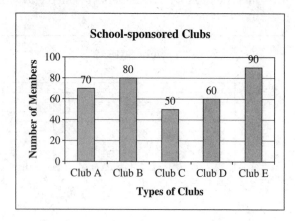

A. 28
B. 32
C. 42
D. 48

Questions 5 and 6 are based on the scatterplot below.

The scatterplot below shows a student's grade on a test against the number of hours that student spent studying each night.

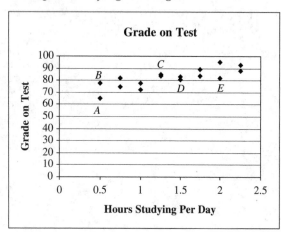

5. Which of the labeled points above represents the highest grade on the test?

A. *A*
B. *B*
C. *C*
D. *D*

6. Based on the correlation of the data, which would most likely be the test grade of a student who studies 3 hours a day?

A. 60
B. 70
C. 80
D. 90

Questions 7 and 8 are based on the line graph below.

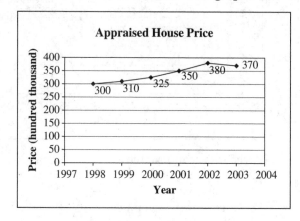

The appraised price of a house is shown in the line graph above.

7. What is the average increase in the price of the house per year?

A. $5,000
B. $10,000
C. $14,000
D. $35,000

8. What is the percent decrease in the price of the house from 2002 to 2003 to the nearest hundredth of a percent? Write your answer without the percent sign. (ex. 10% = 10).

9. Below is a scatter plot showing the number of acres of land a house is on compared to the price of the house. What percentage of the houses on more than 4 acres cost less than $400,000?

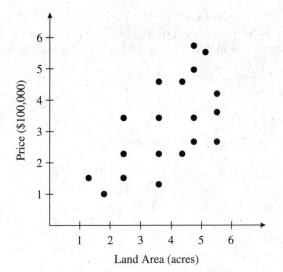

10.

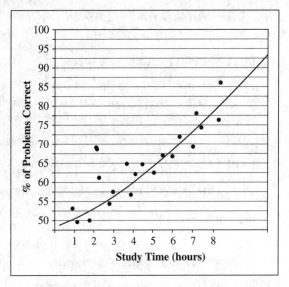

The graph shows the relationship between the percent of problems correct and the number of hours spent studying for 20 students on a recent test. The line of best fit is included. A student whose data is not included above spent 4 hours studying. According to the line of best fit, which of these is closest to his predicted percent of questions correct?

A. 61%

B. 65%

C. 70%

D. 90%

EXPLAINED ANSWERS

1. **EXPLANATION: Choice A is correct.**

   The graph needs to be as close to flat as possible to indicate no change or a small change, and that is clearly taking place between February and March. If you looked for the biggest or smallest drop in cases or the smallest number of cases, rather than the smallest change, you may have selected one of the trap answer choices.

2. **EXPLANATION: Choice D is correct.**

   Divide the change in average points by the average points in 1995.

   Percent increase $= \dfrac{22 - 18}{18} = \dfrac{4}{18} = 22.22\%$.

3. **EXPLANATION: Choice C is correct.**

   First find the total number of people surveyed.

Comedy $= 20$	Suspense $= 30$	Action $= 25$
Horror $= 15$	Drama $= 40$	Other $= 20$

   $20 + 30 + 25 + 15 + 40 + 20 = 150$ were surveyed

   Now find the number of people who chose action or chose horror.

   $25 + 15 = 40$ were either action or horror

   Write a fraction and calculate percent.

   Percent of Action or Horror $= \dfrac{40}{150} = 26\dfrac{2}{3}\%$.

4. **EXPLANATION: Choice C is correct.**

   40% of the students in Club A are in Club B. That means 60% of the students in Club A are not in Club B. Find 60% of 70. $0.60 \times 70 = 42$ students in Club A are not in Club B.

5. **EXPLANATION: Choice C is correct.**

   Test grades are along the vertical axis. That means the higher the point, the better the test grade. Of the indicated points, point C is the highest.

6. **EXPLANATION: Choice D is correct.**

   This scatterplot indicates a positive correlation between study time and grade. Every student who studied more than two hours scored 80 or better on the test; several of these in the 90 range. Therefore, of the choices given, a student who studies 3 hours a day would be expected to earn a 90.

7. **EXPLANATION: Choice C is correct.**

   The house price increased from \$300,000 in 1998 to \$370,000 by 2003. Divide the total increase in price, \$70,000, by the 5 years over which the increase took place.

   The average yearly increase is:

   $\dfrac{370,000 - 300,000}{2003 - 1998} = \dfrac{\$70,000}{5} = \$14,000$.

8. **EXPLANATION: The correct answer is 2.63.**

Find the change in house price and divide by the original house price.

$$\text{Percent decrease} = \frac{380,000 - 370,000}{380,000} = 0.0263 = 2.63\%.$$

9. **EXPLANATION: The correct answer is 50.**

There are 10 houses that have more than four acres. Five of these houses cost less than \$400,000. Therefore, 50% of the houses with more than 4 acres of land cost less than \$400,000.

10. **EXPLANATION: Choice A is correct.**

Looking at the 4-hour spot on the x-axis, it intersects with the line of best fit just above 60% correct. Wrong answers might have resulted from not being sure how to interpret the line of best fit and instead using the dots at the 4-hour study-time mark.

CHAPTER 21

SAMPLING AND STATISCAL INFERENCE

There are two main types of SAT problems dealing with Sampling and Statistical Inference. You will be asked questions about appropriate conclusions that can be made from a sample. You will be also asked what is wrong with the sample, the bias, which prevents a proper conclusion from being made. Begin with the mathematics review and then complete and correct the practice problems. There are 2 Solved SAT Problems and 10 Practice SAT Questions with answer explanations.

Inference means drawing conclusions about a larger group, the population, from a smaller included group, the sample.

Population A population is the larger group under study.

Sample A sample is a group chosen from within the population. A sample should be unbiased, meaning that it accurately represents the population.

Random Sample A random sample is a sample chosen randomly (unbiased sample) from the population. In other words, each member of the sample has the same probability of being chosen as any other member. SAT questions are typically based on a random sample.

Margin of Error A margin of error is the percent of error in results of a survey based on a random sample. A margin of error of 5% means the most likely result on the population is 5% of the survey result. So, if a survey result based on random sample of 100 people was 20% of the sample it is likely that the result in the population would be 15% to 25%.

Confidence Level The confidence level is the percent probability that the population result will be in the margin of error. In the example above, a 95% confidence level means there is a 95% likelihood that the result in the population will be in the margin of error. You will not use the confidence level reported in SAT questions to solve the problem. Just remember that it is likely but not certain the results for the population will be in margin of error.

Answering Inference Questions

Inference questions usually ask you to draw a conclusion based on the reported results, often using the margin of error. Some inference questions ask about the validity of conclusions. Other questions include answer choices that are traps. That is these choices represents conclusions that seem to be appropriate but are not. For example, the sample may be based on one population and the conclusion is about a different population.

Example 1.

School Food Services conducted a survey. The high school has 1200 students who eat lunch in the school cafeteria. The survey was distributed to and collected from a random sample of 400 students who ate lunch in the cafeteria. The survey asked students to choose from one among five lunch choices.

The survey results, with a 4% margin of error at a 95% confidence level, are given below.

Pizza	150
Chicken	100
Green Salad	46
Fruit Salad	54
Macaroni and Cheese	50

Which of the following conclusions can be drawn from these survey results?

1. Based on the sample results, 95% of all the students who eat lunch in the cafeteria are likely to prefer a salad.

    Incorrect. The 95% refers to the confidence level and not to anything about the sample results or the margin of error.

2. Based on the sample results, it is most likely that of the students who eat lunch in the cafeteria about 27% are likely to prefer a salad.

    Incorrect. The salads together are 100 preferences out of 400 or 25%. The margin of error is 4%, which means a result is likely to be 25% 4%, or 21% to 29%.

3. Based on the sample results, it is most likely that about 50 students who eat lunch in the cafeteria will prefer macaroni and cheese.

    Incorrect. The number 50 is the number of students from the sample who choose macaroni and cheese, not the total number of students who are likely to prefer macaroni and cheese.

    Follow these steps to find it is most likely that about 8.5% to 16.5% students who eat lunch in the cafeteria will prefer macaroni and cheese.

    The macaroni and cheese preferences are 50 preferences out of 400 or 12.5%. The margin of error is 4%, which means a result is likely to be 12.5% 4%, or 8.5% to 16.5%.

4. Based on the sample results, it is most likely that of the students in the school that from 46% to 54% students will choose pizza or macaroni and cheese.

    Incorrect. Only students who ate lunch in the school were sampled so the sample results cannot be extended to all the students in the school. The conclusion could be drawn if referredjust to students who ate lunch in the cafeteria.

5. Based on the sample results the School Food Services can be sure that that of the students who eat lunch in the cafeteria about 58.5% to 66.5% are likely to prefer a chicken or pizza.

    Incorrect. The staff cannot be <u>sure</u> the percent who prefer chicken or pizza. The percent who prefer chicken or pizza is 95% likely to be between 58.5% to 66.5%; however, it could be higher or lower.

    This answer would be correct if the conclusion stated that it would be most likely the percent who each lunch in the cafeteria prefer chicken or pizza.

6. The survey used the same random sample to ask, "Would you pay extra for table service in the cafeteria?" Just 30 students indicated they would pay for this service. Based on this response rate with a 4% margin of error, how many students who eat lunch in the cafeteria are likely to pay for table service.

Thirty students out of 400 is 7.5%. The margin of error is 4%, which means a result is likely to be 7.5% 4%, or 3.5% to 11.5%

Now find the number of students.

1200 students eat in the cafeteria. 3.5% of 12000 is 42. 11.5% of 1200 is 138. That means it is likely that 42 to 138 students who eat lunch in the cafeteria will pay for table service.

Example 2.

A large company has employees at three locations, Alpha, Beta, and Cero. A research study is in the planning stages to accomplish these goals.

(1) Estimate the present of employees at location Alpha, where there is no current remote work, who want to work remotely,

(2) Determine exactly the percent of employees at location Beta, which already has remote workers.

(3) Apply the results for Location Alpha to Location Cero.

Briefly describe the experimental designs that would accomplish each of these three goals.

(1) Estimate the percent of workers at Location Alpha who want to work remotely. Ask for and receive responses from a random sample of Alpha workers asking whether the worker want to work remotely. Find the percent or workers from the sample who want to work remotely. Find the percent margin of error and apply that margin of error to the number of workers at Location Alpha.

(2) Determine exactly the percent of workers at Location Beta who are working remotely. Experiments produce estimates with uncertainty so there is no appropriate experimental design that will produce an exact result. An actual count is required to determine <u>exactly</u> how many workers at Location Beta are working remotely.

(3) Apply the results from Location Alpha to Location Cero. The results from one location cannot be applied to another location because a sample from one population cannot be used to draw conclusions about another population.

Practice Questions

1. The city council wants to know people's feelings on a plan to build a new school. Would the best sample for this survey be a (A) randomly selected group of parents who would use the new school, (B) a randomly selected group of citizens, or (C) a randomly selected group of people who voted in the last election?

2. A candy manufacturer took a random sample of 100 bars, including 10 cracked bars, from a manufacturing run of 1000 bars. At this rate, how many bars in the manufacturing run will be cracked.

3. What would be an appropriate way to determine whether students at an elementary school would be motivated by a longer recess period?

Practice Answers

1. A randomly selected group of citizens.
2. 100.
3. Surveying a random group of students of all ages that is representative of the size of the school. (Note, however, that this isn't a great survey: different age groups might need different recess times.)

SOLVED SAT PROBLEMS

1. A school district has 15 elementary schools located around the city. The superintendent thinks that parent approval with the schools is similar across all of the elementary schools. Which of the sampling methods listed is the best way for the superintendent to determine the proportion of parents who approve of the elementary schools?

    A. Randomly picking one of the schools and doing a survey of all the parents at that school

    B. Randomly picking 10 parents from each school and surveying them

    C. Surveying the 10 parents who are most involved at each school and 10 who are not at all involved in the life of the school

    D. Creating a survey form online in which parents can give their thoughts and using the first 100 responses

    EXPLANATION: Choice B is correct.

    Look at each answer choice in turn. If the superintendent picks only one school, as in choice A, he is likely to get opinions that might be skewed—those in a different part of the city might have very different ideas. Because the question specifies that the district has many schools and that they're all over the city, this wouldn't lead to a useful survey. Choice C is likely to get skewed responses as well, since very involved parents are likely to be happy with the school and those who are less involved, if they take the survey at all, are likely to be less happy. Finally, those with strong opinions are most likely to answer an online survey, so the first 100 responses as specified in D are likely to each be strongly either happy or unhappy: again, not a way to get a representative sample. The best option, therefore, is choice B, because it uses all schools and randomizes which parents at each answer the survey.

2. In order to figure out the mean number of items purchased made by the shoppers at a department store opening, the store owner surveyed a group of young women who were waiting in line for a fitting room. For the 30 shoppers surveyed, the mean number of items purchased was 3.1. Which of these is a true statement?

   A. The mean number of purchases by all shoppers in the store that day was 3.1.

   B. No conclusion about the mean number of purchases can be drawn because the sample size is inappropriately small.

   C. The method of the sample is flawed and might give a biased estimate of the number of purchases made by all shoppers.

   D. The sampling method is appropriate and is likely to give a good estimate of the number of purchases made by all shoppers.

   EXPLANATION: Choice C is correct.

   Since we don't know how many shoppers were in the store, it's possible that choice B is true, but not definite. Remember that the question is asking for what "must" be true. Choice A, similarly, is possible but isn't a conclusion we can safely draw. Note that choices D and C are opposites: because both cannot be true, it is likely that one is the answer. In this case, choice C, which says that it "may" be a biased sample, is the best answer. The people in line might be buying more or less than other shoppers of different demographics or at different times of day, and that would lead to the issues mentioned in the choice.

SAMPLING AND STATISTICAL INFERENCE
PRACTICE SAT QUESTIONS

ANSWER SHEET

Choose the correct answer.

1. (A) (B) (C) (D)
2. (A) (B) (C) (D)
3. (A) (B) (C) (D)
4. (A) (B) (C) (D)
5. (A) (B) (C) (D)
6. (A) (B) (C) (D)
7. (A) (B) (C) (D)
8. (A) (B) (C) (D)
9. (A) (B) (C) (D)
10. (A) (B) (C) (D)

PRACTICE SAT QUESTIONS

1. A large car dealership hired a research firm to survey customers to determine the percent of dealership visitors who lived outside the town where the dealership is located. The firm randomly sampled visitors for one month and found that 12% of those sampled lived outside the town, with a 2% margin of error. Based on the survey results with that margin of error, which of the following choices is the most reasonable conclusion about all visitors to the dealership that month?

   A. It is reasonable to conclude that more than 12% of the dealership visitors that month were from outside the town where the dealership was located.

   B. It is reasonable to conclude that from 10% to 14% of the dealership's visitors that month were from outside the town where the dealership was located.

   C. It is reasonable to conclude that from 12% to 14% of the dealership's visitors for that month were from outside the town where the dealership was located.

   D. It is reasonable to conclude that the dealership can be certain that from 10% to 14% of the dealership's visitors that month were from outside the town where the dealership was located.

2. To decide whether Curriculum A was useful in teaching remedial reading, a school district conducted a study in which 500 randomly chosen students out of a large population of students with low reading scores were chosen. 250 of them were randomly assigned Curriculum A, whereas the other 250 did not get any additional instruction. The results demonstrated that those who received Curriculum A had significantly increased reading scores compared to those who did not. Which of the following can be most reasonably concluded from this information?

   A. Curriculum A is the best curriculum for remedial readers.

   B. Curriculum A will improve the reading score of any student who is taught this way.

   C. Curriculum A will improve reading scores.

   D. Curriculum A will most likely improve the reading scores of remedial readers.

3. A city recently surveyed 2,000 randomly selected registered voters, asking each of them "Did you vote in the last municipal election?" All surveys were returned, and of the survey respondents, 34 percent said that they did. Which of the following must be true?

   A. Of all citizens in the city, 34% voted in the last municipal election.

   B. If another 2,000 registered voters were sampled, about 34% of them would report having voted in the last municipal election.

   C. 34% of the voters care about municipal elections.

   D. Of all registered voters in the city, about 66% did not vote in the last municipal election.

4. A researcher chose a random sample of 300 bicycle racers who liked the design of a particular racing bike. Then the same 300 bicycle racers actually rode the bike and were asked if they liked riding the bike or did not like riding the bike. Out of the 300 bicycle racers, 118 also liked riding the bike. Which of the following inferences can appropriately be drawn from this survey result?

   A. Bicycle racers who like the design of this particular racing bike will not like to ride that bike.

   B. Bicycle racers who like the design of a racing bike will not like to ride the bike.

   C. Bicycle racers who like the design of this particular racing bike will usually not like to ride that bike.

   D. Bicycle racers who like the design of a racing bike will usually not like to ride that bike.

5. A study was done of the heights of a certain type of tree in a national park in Kenya. A random sample of trees were measured. The sample contained 20 trees, and $\frac{1}{4}$ were over 30 feet tall. Which of these conclusions can be properly drawn?

   A. Roughly 25% of all the trees in the park are under 30 feet tall.

   B. The majority of the trees in the park are less than 30 feet tall.

   C. The average height of the trees in the study is less than 30 feet.

   D. Roughly 25% of all the trees in the park are over 30 feet tall.

6. A paint company picked 300 customers who had reported they would be interested in buying paint. These people were shown a new potential lipstick color, but of those surveyed, 85% said they would not buy the color. Which of the following can be reasonably concluded from this result?

   A. Many people who are interested in buying paint won't buy this color.

   B. Most people who are interested in buying paint will buy a similar color from another brand.

   C. At least 85% of people who buy paint will not buy this color, but they might consider it if the lipstick is on sale.

   D. More than 85% of the paint company customers will not buy this color.

7. In a recent election, citizens could vote early or on election day one out of two candidates for the same office. The candidate who got more than 50% of the vote won. About 15% of the citizens voted, and 20% of the votes were cast early. Candidate 1 earned 40% of the votes cast early and 30% of the votes on election day. Which of the following can be concluded?

   A. Those voting early were more likely than those on election day to vote for Candidate 1.

   B. If all of those who voted had voted early instead of on election day, Candidate 1 would be the winner of the election.

   C. Citizens voting early were more likely to be hourly workers than those voting on election day.

   D. If all citizens had voted, Candidate 1 would be the winner.

8. To determine the number of pets per household in a town, the director of the town's Animal Advocacy Center surveyed 10 people with pets at a town park. Of the 10 people the director surveyed, the mean number of pets per household was 3.5. Which of these is definitely true?

   A. The mean number of pets in the town is around 3.5 per household.

   B. There are 3.5 pets per household among people interviewed in the park.

   C. There are 3.5 pets per household in the town when the survey was conducted.

   D. The mean number of pets among pet-owning households in the town is about 3.5

9. After a PTO meeting, the principal asked those present to respond anonymously to a poll that asked if the school should implement a different grading policy to allow for extra points for honors classes. The next day in an email blast to the school community, he reported that 23% responded that the school should keep its current grading policy, and 61% responded that they want a change. Which option best encapsulates the issues with the principal using this survey to drive policy?

   A. The percentages don't add up to 100. Thus, the principal cannot draw any conclusions.

   B. The responses to the survey do not represent a random sample of the school population.

   C. That more people were in favor of the change than against it means that the responses were skewed.

   D. The principal should have had the meeting attendees do the poll before the meeting.

10. A camp director asked a random sample of campers how often they thought about camp during the school year. She found that 50% of the campers in the sample thought about camp at least twice a month during the school year. The margin of error for her estimate is 5%. Which conclusion can be drawn?

    A. Most of the campers probably do not think about camp often.

    B. Definitely no more than 55%, but at least 50%, of the campers think about camp at least twice a month, but some think about it more.

    C. The director is between 45% and 55% sure that typical campers think about camp at least twice a month.

    D. It is fairly likely that the percentage of campers who think about camp at least twice a month is between 45% and 55%.

EXPLAINED ANSWERS

1. **EXPLANATION: Choice B is correct.**

   The survey found that 12% of the randomly sampled visitors lived outside the town, The margin of error is 2%, which means the number of visitors who lived outside the town is likely to be 12% ± 2%, or 10% to 14%. Choice (A) is incorrect because it does not incorporate the margin of error, while Choice (C) is incorrect because only the positive margin of error is applied. Choice (D) is incorrect because there is always uncertainty, and there is no guarantee that the result will be in the margin of error.

2. **EXPLANATION: Choice D is correct.**

   Choice A is incorrect because Curriculum A isn't compared, as far as the information tells us, to any other curricula. We only know from the data how it impacts people with low reading scores, not all students, so that rules out choice B. Finally, choice C is too vague—it doesn't specify the population in which this will occur or what "improve" means.

3. **EXPLANATION: Choice D is correct.**

   Since 34% of the sample voted in the previous municipal election, 66% did not. The phrase "about 66%" makes the answer correct because it is not an exact percent—but rather within some margin of error. Choice (A) is incorrect because the sample is from citizens who were registered voters, and this choice mentions all citizens. Choice (B) is incorrect because this second group of registered voters was not randomly sampled. Choice (C) is incorrect because the sample was not based on voters who cared.

4. **EXPLANATION: Choice C is correct.**

   In the sample of 300 bicycle racers who like a particular bike's design, 118, less than half, liked riding the bike. So most bicycle racers who liked the design of this particular bike usually did not like to ride the bike. Choice (A) is incorrect because some bicycle racers who liked the design did like to ride the bike. Choices (B) and (D) are incorrect because they refer to racing bikes in general and not to this particular racing bike.

5. **EXPLANATION: Choice D is correct.**

   Choice D is the narrowest conclusion that can be drawn. You cannot make a definite statement from a sample nor can you extrapolate the average, as in choice C, since the information doesn't include specific numbers. Choice A stems from a misreading or math error.

6. **EXPLANATION: Choice A is correct.**

   Note that it is the smallest possible inference to make—often a sign of a correct response on these types of questions. Choices C and D use the terms "at least" and "more than," which isn't supported by the data, and we have no information about what else these customers might buy, as would be needed to make the inference in choice B.

7. **EXPLANATION: Choice A is correct.**

   This can be drawn directly from the data. We can't draw conclusions about the sample, as in choice B, for sure even if they make sense, nor can we predict what they would have done if they had voted at another time, as in choice C: too many factors might intervene to change the outcome. Furthermore, we can't extrapolate from such a small percentage of voters how everyone would feel.

8. **EXPLANATION: Choice B is correct.**

   We know this choice is correct because the question gives this information about the people with pets who were surveyed in the park. Choices (A), (C), and (D) are incorrect, at least, because the sample is not random, and no conclusions can be drawn.

9. **EXPLANATION: Choice B is correct.**

Because those who responded to the poll were the people already at the meeting, they were likely to be more involved members of the school community, and do not necessarily accurately represent the school community as a whole.

10. **EXPLANATION: Choice D is correct.**

Margin of error is not exact, which explains the wording "fairly likely" in the choice. The margin of error means that the percentage found in the responses may represent a range, but not only in one direction, as in choice B, nor does it have to do with certainty, as in C.

CHAPTER 22

ANGLE RELATIONSHIPS

When lines intersect, a variety of angle types are formed. These angles often have a sum of 90° or a sum of 180°. Many may have the same angle measure. It is important to know which rules apply to which angles. Begin with the mathematics review and then complete and correct the practice problems. There are 2 Solved SAT Problems and 11 Practice SAT Questions with answer explanations.

The diagram below shows two lines, *l* and *m*, intersected by a transversal, *t*.

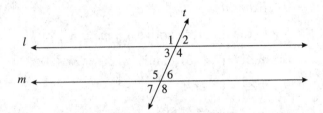

Vertical angles are congruent because their measures are equal.

∠1 and ∠4, ∠2 and ∠3, ∠5 and ∠8, ∠6 and ∠7 are vertical angles.

$\angle 1 \cong \angle 4 \Rightarrow m\angle 1 = m\angle 4$ and $\angle 2 \cong \angle 3 \Rightarrow m\angle 2 = m\angle 3$.

Linear pairs are supplementary because the sum of their measures is 180°.

∠1 and ∠2, ∠2 and ∠4, ∠4 and ∠3, ∠3 and ∠1, ∠5 and ∠6, ∠6 and ∠8, ∠8 and ∠7, ∠7 and ∠5 are linear pairs.

$m\angle 1 + m\angle 2 = 180°$ and $m\angle 2 + m\angle 4 = 180°$ and $m\angle 4 + m\angle 3 = 180°$ and $m\angle 3 + m\angle 1 = 180°$.

$m\angle 5 + m\angle 6 = 180°$ and $m\angle 6 + m\angle 8 = 180°$ and $m\angle 8 + m\angle 7 = 180°$ and $m\angle 7 + m\angle 5 = 180°$.

Alternate interior angles are congruent if *l* is parallel to *m*, *l* ∥ *m*.

∠3 and ∠6, ∠4 and ∠5 are alternate interior angles.

Alternate exterior angles are congruent if *l* is parallel to *m*, *l* ∥ *m*.

∠2 and ∠7, ∠1 and ∠8 are alternate exterior angles.

Corresponding angles are congruent if *l* is parallel to *m*, *l* ∥ *m*.

∠1 and ∠5, ∠2 and ∠6, ∠3 and ∠7, ∠4 and ∠8 are corresponding angles.

Same side interior angles are supplementary if *l* is parallel to *m*, *l* ∥ *m*.

∠3 and ∠5, ∠4 and ∠6 are side side interior angles.

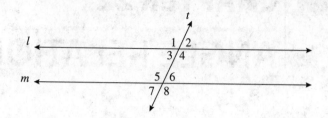

Example 1.

In the diagram above, $m\angle 2 = 30°$ and $l \parallel m$. Find the measure of all the other angles.

$m\angle 1 = m\angle 4 = m\angle 5 = m\angle 8 = 150°$.

$m\angle 2 = m\angle 3 = m\angle 6 = m\angle 7 = 30°$.

Example 2.

In the diagram above, $l \parallel m$.

$m\angle 1 = (7x - 3)$ and $m\angle 6 = (3x + 13)$.

$m\angle 1 = ?$ and $m\angle 6 = ?$

$m\angle 1 + m\angle 6 = 180°$.

Substitute $m\angle 1 = 7x - 3$ and $m\angle 6 = 3x + 13$. $7x - 3 + 3x + 13 = 180$.

Solve for x. $10x + 10 = 180 \Rightarrow 10x = 170 \Rightarrow x = 17$.

Substitute $x = 17$ for $m\angle 1$ and $m\angle 6$.

$m\angle 1 = 7(17) - 3 = 116°$ and $m\angle 6 = 3(17) + 3 = 64°$.

In the diagram below, line r is perpendicular to line s, written $r \perp s$. Perpendicular lines form right angles $m\angle AOB = m\angle BOD = m\angle DOC = m\angle COA = 90°$.

$\angle 1$ and $\angle 2$ are referred to as complimentary angles because $m\angle 1 + m\angle 2 = 90°$.

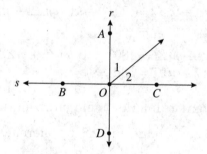

Example 3.

In the diagram above, $r \perp s$, $m\angle 1 = (4x - 9)°$, $m\angle 2 = (2x + 3)°$. $m\angle 1 = ?$ and $m\angle 2 = ?$

Find the measure of angle 1 and angle 2.

$m\angle 1 + m\angle 2 = 90° \Rightarrow 4x - 9 + 2x + 3 = 90$.

Solve for x. $6x - 6 = 90$, $6x = 96 \Rightarrow x = 16$.

Substitute $x = 16$ for both $m\angle 1$ and $m\angle 2$.

$m\angle 1 = 4(16) - 9 = 55°$ and $m\angle 2 = 2(16) + 3 = 35°$.

The sum of the interior angles of a triangle is 180°.

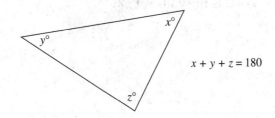

$x + y + z = 180$

Example 4.

What is the value of x?
Sum of the angle measures equals 180.

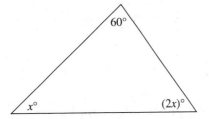

$x + 2x + 60 = 180$

Solve for x. $3x + 60 = 180$.

$3x = 120 \Rightarrow x = 40$.

The measure of the exterior angle of a triangle is equal to the sum of the two remote interior angles. The remote interior angles are the two angles most distant from the exterior angle.

$z = x + y$

Example 5.

What is the value of x?

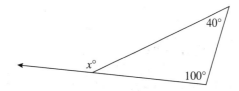

$x = 100 + 40 = 140$.

Example 6.

What is the value of x?

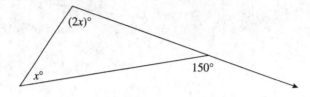

$x + 2x = 150 \Rightarrow 3x = 150 \Rightarrow x = 50.$

Parallelogram—a quadrilateral with:

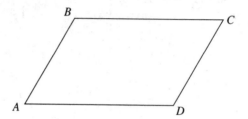

1. Opposite sides parallel. $\overline{AB} \parallel \overline{CD}$ and $\overline{BC} \parallel \overline{AD}$
2. Opposite sides congruent. $\overline{AB} \cong \overline{CD}$ and $\overline{BC} \cong \overline{AD}$
3. Opposite angles congruent. $\angle B \cong \angle D$ and $\angle A \cong \angle C$
4. Consecutive angles supplementary.

$m\angle A + m\angle B = 180°$ and $m\angle B + m\angle C = 180°$.

$m\angle C + m\angle D = 180°$ and $m\angle D + m\angle A = 180°$.

Example 7.

In the parallelogram seen below, $m\angle ABC =$

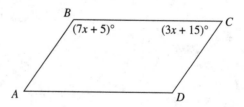

$m\angle B + m\angle C = 180°$

$7x + 5 + 3x + 15 = 180° \Rightarrow 10x + 20 = 180°$

$10x = 160° \Rightarrow x = 16$

$m\angle ABC \angle$ Substitute 16 for x. $7(x) + 5 = 7(16) + 5 = 117°$.

Every polygon can be partitioned into nonoverlapping triangles as shown below.

 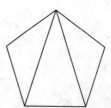 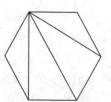

The number of triangles that are formed is two less than the number of sides. The sum of the measures of the angles in a triangle is 180°. Therefore, any polygon having n sides has an angle sum of $(n-2) \times 180°$.

Regular polygon—a polygon where all sides are congruent, and all angles are congruent.

A central angle of a regular polygon is formed when segments are constructed from the center of the polygon to the vertices of the polygon.

Each central angle of a regular polygon is congruent, having a measure of $\dfrac{360°}{n}$ where n is the number of angles. The triangles that are formed through this process are also congruent.

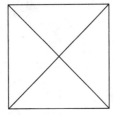 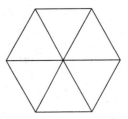

Example 8.

What is the angle sum of an octagon?

An octagon has eight sides.

$(8 - 2) \times 180° = 6 \times 180° = 1,080°$.

Example 9.

$m\angle A + m\angle B + m\angle C =$

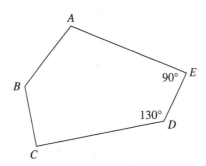

$m\angle A + m\angle B + m\angle C + m\angle D + m\angle E = 3(180°) = 540°$

Substitute $m\angle D = 130°$ $m\angle E = 90°$.

$m\angle A + m\angle B + m\angle C + 130° + 90° = 540°$

$m\angle A + m\angle B + m\angle C + 220° = 540°$

$m\angle A + m\angle B + m\angle C = 320°$

Example 10.

What is the measure of each central angle in a regular hexagon?
Calculate 360° divided by the number of sides.

$\dfrac{360°}{6} = 60°$

Practice Questions

Use the diagram below for problems 1 and 2 given that $l \parallel m$.

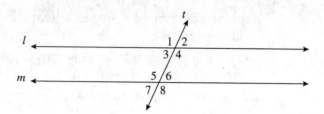

1. $m\angle 5 = (4x + 6)°$ and $m\angle 8 = (7x - 15)$.
   $m\angle 5 = ?$ and $m\angle 8 = ?$

2. $m\angle 4 = (3x + 5)°$ and $m\angle 6 = (4x - 7)°$
   $m\angle 4 = ?$ and $m\angle 6 = ?$

3. What is the value of x?

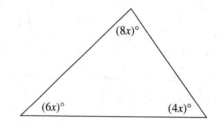

4. What is the value of x?

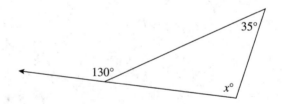

Practice Answers

1. $m\angle 5 = 34°$ and $m\angle 8 = 34°$
2. $m\angle 4 = 83°$ and $m\angle 6 = 97°$
3. $x = 10$
4. $x = 95°$

SAT SOLVED PROBLEMS

1. In the diagram shown below, find the value of x given that line p is parallel to line q.

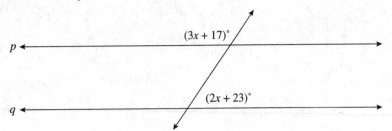

A. $x = 6$

B. $x = 9$

C. $x = 28$

D. $x = 40$

EXPLANATION: Choice C is correct.

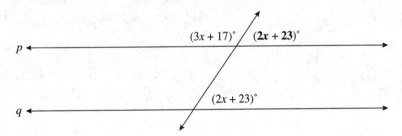

The additional $(2x + 23)°$ was placed in the picture because corresponding angles are congruent. Now from the new picture, the two angles represented by $(3x + 17)°$ and $(2x + 23)°$ are a linear pair. Therefore $(3x + 17) + (2x + 23) = 180$.

Simplify left sides of the equation: $5x + 40 = 180$

Subtract 40 from both sides of the equation: $5x = 140$

Divide by 5 on both sides of the equation: $x = 28$

2. The diagram below shows intersecting lines p, q, and r. Find the value of x. Note: figure not drawn to scale.

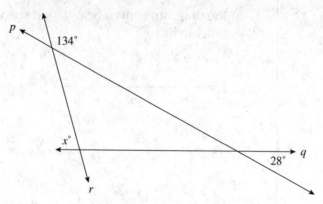

EXPLANATION: The correct answer is 74.

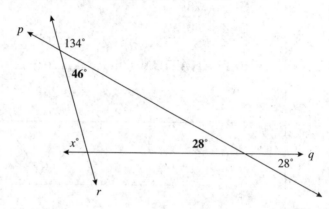

The 46° was added to the picture because linear pairs are supplementary.

The 28° was added to the picture because vertical angles are congruent.

The intersecting lines create a triangle and $x°$ is the measure of an exterior angle.

46° and 28° are the measure of the two remote interior angles.

Therefore, $x = 46 + 28 = 74$.

ANGLE RELATIONSHIPS
PRACTICE SAT QUESTIONS

ANSWER SHEET

Choose the correct answer.
If no choices are given, grid the answers in the section at the bottom of the page.

1. Ⓐ Ⓑ Ⓒ Ⓓ   11. Ⓐ Ⓑ Ⓒ Ⓓ
2. Ⓐ Ⓑ Ⓒ Ⓓ
3. Ⓐ Ⓑ Ⓒ Ⓓ
4. Ⓐ Ⓑ Ⓒ Ⓓ
5. Ⓐ Ⓑ Ⓒ Ⓓ
6. Ⓐ Ⓑ Ⓒ Ⓓ
7. GRID
8. GRID
9. GRID
10. Ⓐ Ⓑ Ⓒ Ⓓ

Use the answer spaces in the grids below if the question requires a grid-in response.

Student-Produced Responses ONLY ANSWERS ENTERED IN THE CIRCLES IN EACH GRID WILL BE SCORED. YOU WILL NOT RECEIVE CREDIT FOR ANYTHING WRITTEN IN THE BOXES ABOVE THE CIRCLES.

7.

8.

9.

PRACTICE SAT QUESTIONS

1. Which of the following statements is correct?

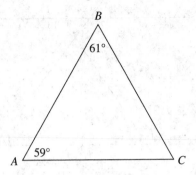

A. $AB < BC < AC$
B. $BC < AB < AC$
C. $AB < AC < BC$
D. $AC < BC < AB$

2. $m\angle CBA =$

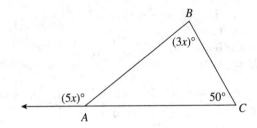

A. 125
B. 105
C. 75
D. 50

3. What is the value of x?

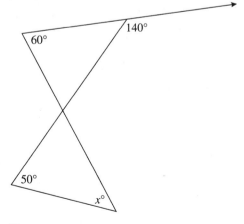

A. 40
B. 50
C. 60
D. 70

4. In the parallelogram below, if x is 4 times as big as y, then $x - y =$

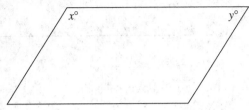

A. 36
B. 108
C. 144
D. 180

5.

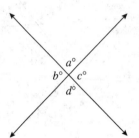

Which of the following statements must be true?

   I. $a = d$
  II. $a + c = 180$
 III. $a + d = 90$

A. I
B. II
C. I and II
D. I and III

6.

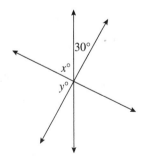

$x + y =$
A. 90
B. 110
C. 120
D. 150

7.

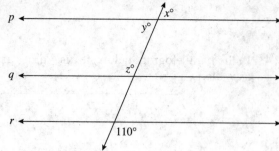

In the figure above, $p \parallel q \parallel r$.

$x + y + z =$

8. $x + y =$

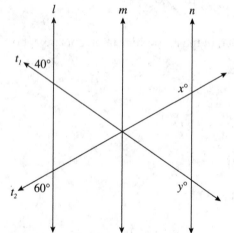

9.

In the figure above, $l \parallel m \parallel n$.

$x + y =$

10.

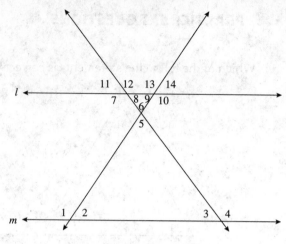

In the figure above, $l \parallel m$.

Which of the following pairs of angles are supplementary?

 I. 1 and 9

 II. 6 and 8

 III. 13 and 14

A. II
B. I and II
C. II and III
D. I and III

11. The number of radians in a 540-degree angle can be written as $m\pi$. What is m?

A. $\dfrac{1}{2}$

B. 1
C. 2
D. 3

EXPLAINED ANSWERS

1. **EXPLANATION: Choice B is correct.**

   The sum of the angles is 180°, so $m\angle C = 60°$.

   $m\angle A < m\angle C < m\angle B \Rightarrow BC < AB < AC$.

2. **EXPLANATION: Choice C is correct.**

   The sum of the exterior angles equals the sum of the remote interior angles.

   $3x + 50 = 5x \Rightarrow 50 = 2x \Rightarrow 25 = x \Rightarrow m\angle CBA$

   Substitute 25 for x. $3x = 3(25) = 75°$.

3. **EXPLANATION: Choice B is correct.**

   The sum of the angles is 180°. Use what you know about linear pairs and vertical angles to find the bold angle measures.

   $x + 50 + 80 = 180 \Rightarrow x + 130 = 180 \Rightarrow x = 50$.

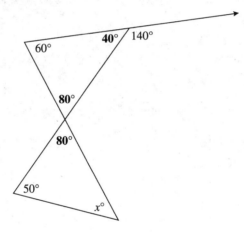

4. **EXPLANATION: Choice B is correct.**

   The sum of consecutive angles is 180°
   $x + y = 180$.
   Substitute $4y$ for x. $4y + y = 180 \Rightarrow 5y = 180$
   Solve for y. $y = 36$.
   Substitute 36 for y. $x = 4(36) = 144$.
   $x - y = 144 - 36 = 108$.

5. **EXPLANATION: Choice C is correct.**

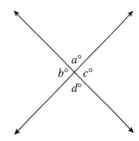

   I. Vertical angles are congruent, so $a = d$.
   II. Linear pairs are supplementary, so $a + c = 180°$.

6. **EXPLANATION: Choice D is correct.**

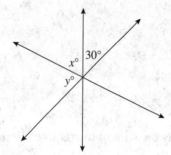

$x + y + 30$ "forms" a straight line, so $x + y + 30 = 180$.
So $x + y = 150$.

7. **EXPLANATION: The correct answer is 250.**

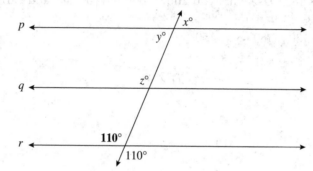

Corresponding angles are congruent, so $z = 110$.
$y + z = 180$.
Same side interior angles are supplementary, so $y + 110 = 180$.
$y = 70$.
Vertical angles are congruent. $x = y \Rightarrow x = 70$.
That means, $x + y + z = 70 + 70 + 110 = 250$.

8. **EXPLANATION: The correct answer is 140.**

The (x) and (y) measures
inside the triangle have
been added to show con-
gruent vertical angles.

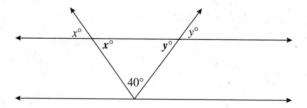

The sum of the angles in a triangle is $180°$.
$x + y + 40 = 180 \Rightarrow x + y = 140$.

9. **EXPLANATION: The correct answer is 260.**

Use the rules for parallel lines cut by a transversal to find the angle measures.

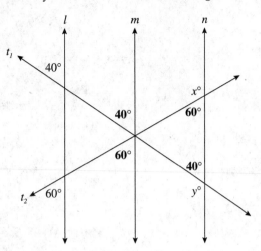

A linear pair is supplementary, so:

$x + 60 = 180$ and $x = 120$.

$y + 40 = 180$ and $y = 140$.

That means, $x + y = 120 + 140 = 260$.

10. **EXPLANATION: Choice D is correct.**

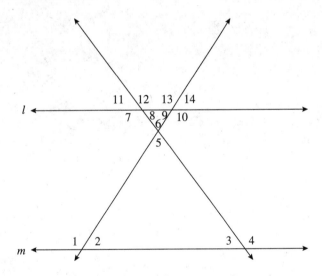

I. YES $\angle 1$ and $\angle 9$ because same side interior angles are supplementary.

II. NO $\angle 6$ and $\angle 8$ are two angles in a triangle. That means the sum of the measures of these angles is less than 180°. They cannot be supplementary.

III. YES $\angle 13$ and $\angle 14$ because a linear pair is supplementary.

11. **EXPLANATION: Choice D is correct.**

To find the number of radians in an angle, multiply by $\frac{\pi}{180}$. $540 \times \frac{\pi}{180}$ = approximately 9.4, or 3π. Thus, $m = 3$.

CHAPTER 23

TRIANGLES, RECTANGLES, AND OTHER POLYGONS

SAT problems focusing on triangles often deal with finding the length of a missing side, which might be accomplished by using a scale factor of similar triangles. When dealing with right triangles, try using rules about special right triangles or the Pythagorean Theorem. Besides just finding side length, you will see area and perimeter questions when working with triangles and rectangles on the SAT. Begin with the mathematics review and then complete and correct the practice problems. There are 2 Solved SAT Problems and 10 Practice SAT Questions with answer explanations.

In **similar triangles**, corresponding angles are congruent and corresponding sides have lengths that are proportional, that is, have the same scale factor, to each other. It is this relationship between side lengths that is useful in finding the lengths of missing sides.

Example 1.

In the figure below, triangle ABC is similar to triangle DEF ($\triangle ABC \sim DEF$)
  Find the length of segment CB and segment DF.

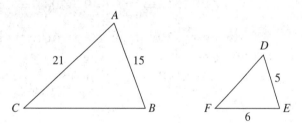

Since $\triangle ABC \sim \triangle DEF$, $\overline{AB}$ and $\overline{DE}$ are matching sides; this tells us the scale factor from $\triangle DEF$ to $\triangle ABC$ is 3. $5(3) = 15$.

Therefore, $CB = 6(3) = 18$ and $DF = 21\left(\dfrac{1}{3}\right) = 7$.

Example 2.

In the figure below, $\overline{LM} \parallel \overline{PQ}$. Find the length of segment LM.

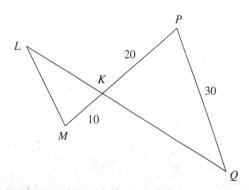

Since $\overline{LM} \parallel \overline{PQ}$, $\angle L \cong \angle Q$ and $\angle M \cong \angle P$ because these are alternate interior angles. Also, $\angle LKM \cong \angle QKP$ because they are vertical angles. Therefore, $\triangle LKM \sim \triangle QKP$.

Since $\triangle LKM \sim \triangle QKP$ and $\overline{PK}$ and $\overline{MK}$ are matching sides, this tells us the scale factor from $\triangle QKP$ to $\triangle LKM$ is $\dfrac{1}{2}$. $20\left(\dfrac{1}{2}\right) = 10$. Therefore, $LM = 30\left(\dfrac{1}{2}\right) = 15$.

The *Pythagorean Theorem* states that in a right triangle, the sum of the square of the legs is equal to the square of the hypotenuse.

$$(\text{leg 1})^2 + (\text{leg 2})^2 = (\text{hypotenuse})$$
$$a^2 \quad + \quad b^2 \quad = \quad c^2.$$

Example 3.

What is the value of x?

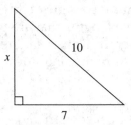

$$x^2 + 7^2 = 10^2 \quad \Rightarrow \quad x^2 + 49 = 100 \quad \Rightarrow \quad x^2 = 51 \quad \Rightarrow \quad x = \sqrt{51}$$

A 3-4-5 right triangle is a right triangle where the ratio of leg 1 : leg 2 : hypotenuse $= 3 : 4 : 5$.

This ratio meets the requirements of the Pythagorean Theorem $3^2 + 4^2 = 9 + 16 = 25 = 5^2$.

Example 4.

What is the value of x?

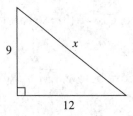

$$3 : 4 : 5 = 9 : 12 : x = (3 \cdot 3) : (3 \cdot 4) : x \Rightarrow x = 3 \cdot 5 = 15$$

Or use the Pythagorean Theorem.

$$9^2 + 12^2 = x^2 \quad \Rightarrow \quad 81 + 144 = 225 = x^2 \quad \Rightarrow \quad 15 = x$$

A 5-12-13 right triangle is a right triangle where the ratio of leg 1 : leg 2 : hypotenuse $= 5 : 12 : 13$.

This ratio meets the requirements of the Pythagorean Theorem $5^2 + 12^2 = 25 + 144 = 169 = 13^2$.

Example 5.

What is the value of x?

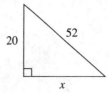

$5 : 12 : 13 = 20 : x : 52 = (4 \cdot 5) : x : (4 \cdot 13)$ so $x = 4 \cdot 12 = 48$

Or use the Pythagorean Theorem.

$x^2 + 20^2 = 52^2 \Rightarrow x^2 + 400 = 2{,}704 \Rightarrow x^2 = 2{,}304 \Rightarrow x = 48$

A 45°-45°-90° right triangle has a special relationship among the sides. If the sides across from the 45° angles are x, the side across from the 90° angle is $x\sqrt{2}$.

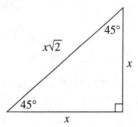

A 30°-60°-90° right triangle also has a special relationship among the sides. If the side across from the 30° angle is x, the side across from the 60° angle is $x\sqrt{3}$ and the side across from the 90° angle is $2x$.

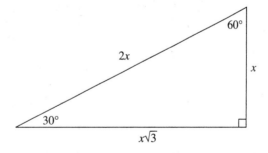

Example 6.

What is the value of x and y?

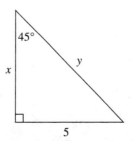

Using 45°-45°-90°, it is clear that $x = 5$, and $y = 5\sqrt{2}$.

Example 7.

What is the value of x and y?

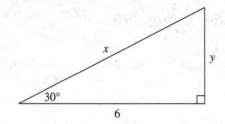

Using $30°\text{-}60°\text{-}90° \Rightarrow y\sqrt{3} = 6 \Rightarrow y = \dfrac{6}{\sqrt{3}}$

$y = \dfrac{6 \times \sqrt{3}}{\sqrt{3} \times \sqrt{3}} = \dfrac{6\sqrt{3}}{3} = 2\sqrt{3}$

Because $y = 2\sqrt{3} \Rightarrow x = 2 \cdot (2\sqrt{3}) = 4\sqrt{3}$

Example 8.

What is the area of the figures below?

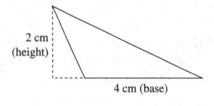

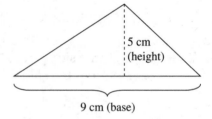

$A = \dfrac{1}{2} \times b \times h$ $A = \dfrac{1}{2} \times b \times h$

$= \dfrac{1}{2} \times 4 \times 2 = \dfrac{1}{2} \times 8 = 4 \text{ cm}^2$ $= \dfrac{1}{2} \times 9 \times 5 = \dfrac{45}{2} = 22.5 \text{ cm}^2$

Example 9.

A triangle has a height of 15 in and an area 45 in². What is the length of the base of the triangle?

$A = \dfrac{1}{2} \times b \times h \quad \Rightarrow \quad 45 = \dfrac{1}{2} \times b \times 15 \quad \Rightarrow \quad 90 = b \times 15 \quad \Rightarrow \quad 6 \text{ in} = b$

The *Triangle Inequality Theorem* states that the sum of the length of any two sides of a triangle is greater than the length of the third side.

Example 10.

What are the possible values of *x*?

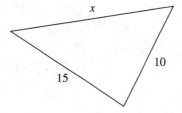

The third side of a triangle has these two characteristics.

1. It is less than the sum of the other two sides, $x < 15 + 10$, $x < 25$.
2. It is greater than the difference of the other two sides $x > 15 - 10$, $x > 5$.

That means $5 < x < 25$.

The largest angle in a triangle is across from the longest side.
The smallest angle in a triangle is across from the shortest side.

Example 11.

List the angles from smallest to largest.

$m\angle C < m\angle A < m\angle B$

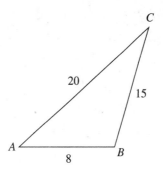

Example 12.

What is the area of the rectangle seen below?

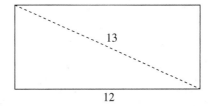

By noticing that a 5-12-13 triangle exists, we can see that the height of the triangle is 5. Therefore, $A = 12 \times 5 = 60$.

Example 13.

The area of the rectangle seen below is 36. What is the value of x?

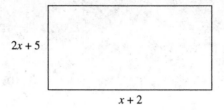

Solve for A = $(2x + 5)(x + 2) = 2x^2 + 9x + 10 = 36$.

Factor $2x^2 + 9x - 26 = 0 \implies (x - 2)(2x + 13) = 0$

$x - 2 = 0 \implies x = 2.\, : 2x + 13 = 0,\ 2x = -13,\ x = -6.5$

$x = 2$ is correct. $x = -6.5$ would create a negative length. That is impossible.

Example 14.

What is the value of x in the square below?

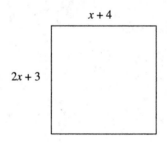

All sides of a square are the same length.

$2x + 3 = x + 4 \implies x = 1.$

Practice Questions

1. In the figure below, triangle ABC is similar to triangle DEF ($\triangle ABC \sim DEF$). Find the perimeter of triangle ABC.

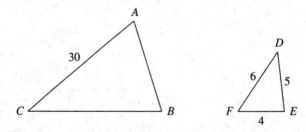

2. What is the value of x?

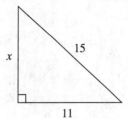

3. What is the value of x?

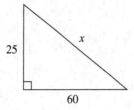

4. What are the values of x and y?

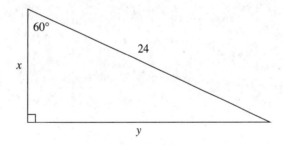

5. What is the area of the figure below?

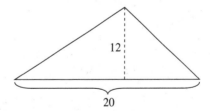

6. A triangle has a base of 12 in and an area 72 in². What is the height of the triangle?

7. What are the possible values of x?

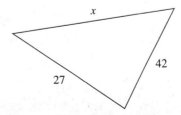

8. List the sides from longest to shortest.

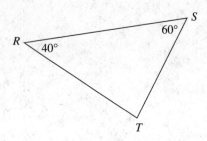

9. The perimeter of the rectangle below is 50. What is the value of x?

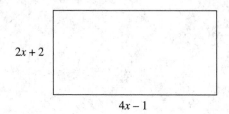

10. What is the perimeter of the square seen below?

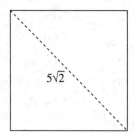

Practice Answers

1. 75.
2. $x = 2\sqrt{26}$.
3. $x = 65$.
4. $x = 12$ and $y = 12\sqrt{3}$.
5. 120.
6. 12.
7. $15 < x < 69$.
8. $RS > RT > ST$.
9. $x = 4$.
10. $P = 20$.

SOLVED SAT PROBLEMS

1. In the figure below, $\overline{TS} \parallel \overline{PQ}$. Find the length of segment SR.

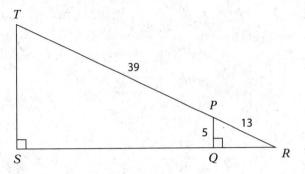

EXPLANATION: The correct answer is 48.

Triangle PQR and triangle TSR are similar triangles. This is because of the parallel lines and the fact that the triangles share an angle.

So $\triangle PQR \sim \triangle TSR$, $\overline{PR}$ and $\overline{TR}$ are corresponding, and $PR = 13$ and $TR = 52$. Therefore, the scale factor from $\triangle PQR$ to $\triangle TSR$ is 4. $13(4) = 52$.

Find the length of $\overline{QR}$. You could use the Pythagorean Theorem.

First place each value in correct position: $\qquad (QR)^2 + 5^2 = 13^2$

Square the numbers: $\qquad\qquad\qquad\qquad (QR)^2 + 25 = 169$

Subtract 25: $\qquad\qquad\qquad\qquad\qquad (QR)^2 = 144$

Take square root of both sides: $\qquad\qquad QR = 12$

You could have also noticed that $QR = 12$ because triangle PQR is a 5-12-13 right triangle.

Since $\overline{QR}$ and $\overline{SR}$ are corresponding and $QR = 12 \rightarrow SR = 12(4) = 48$.

2. Find the area of triangle *JKL* shown below.

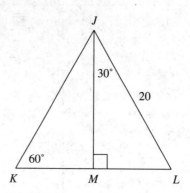

A. $10\sqrt{3}$
B. $20\sqrt{3}$
C. $100\sqrt{3}$
D. $200\sqrt{3}$

EXPLANATION: Choice C is correct.

Based on the given information, triangle *JKL* is an equilateral triangle that has been split into two 30°-60°-90° right triangles. Using rules of 30°-60°-90°, the base and the height of triangle *JKL* can be calculated.

$$2(LM) = JL \quad \rightarrow \quad 2(LM) = 20 \quad \rightarrow \quad LM = 10$$
$$KM = LM = 10$$

Therefore, the base is 20.

$$JM = (LM)\sqrt{3} \quad \rightarrow \quad JM = 10\sqrt{3}$$

Therefore, the height is $10\sqrt{3}$.

So, the area of triangle *JKL* is $\frac{1}{2}(20)(10\sqrt{3}) = (10)(10\sqrt{3}) = 100\sqrt{3}$.

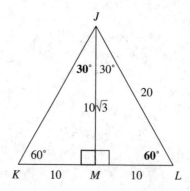

TRIANGLES, RECTANGLES, AND OTHER POLYGONS PRACTICE SAT QUESTIONS

ANSWER SHEET

Choose the correct answer.

1. Ⓐ Ⓑ Ⓒ Ⓓ
2. Ⓐ Ⓑ Ⓒ Ⓓ
3. Ⓐ Ⓑ Ⓒ Ⓓ
4. Ⓐ Ⓑ Ⓒ Ⓓ
5. Ⓐ Ⓑ Ⓒ Ⓓ
6. Ⓐ Ⓑ Ⓒ Ⓓ
7. Ⓐ Ⓑ Ⓒ Ⓓ
8. Ⓐ Ⓑ Ⓒ Ⓓ
9. Ⓐ Ⓑ Ⓒ Ⓓ
10. Ⓐ Ⓑ Ⓒ Ⓓ

PRACTICE SAT QUESTIONS

1. $y - x =$

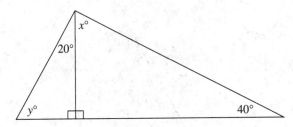

   A. 50
   B. 40
   C. 30
   D. 20

2. What is the area of the triangle below?

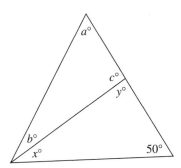

   A. 4 in²
   B. 5 in²
   C. 6 in²
   D. 7 in²

3. $a + b + c + x + y =$

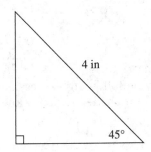

   A. 360
   B. 310
   C. 270
   D. 180

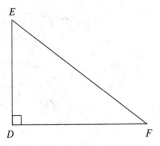

4. In the picture above, triangle *ABC* is similar to triangle *DEF*. What fraction given here expresses tan $< DEF$?

   A. $\dfrac{3}{5}$

   B. $\dfrac{4}{5}$

   C. $\dfrac{4}{3}$

   D. $\dfrac{5}{3}$

5. If the area of the rectangle is 120, what is the area of triangle *CPD*?

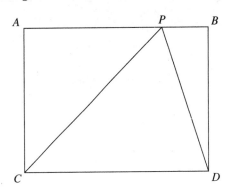

   A. 60
   B. 100
   C. 120
   D. 200

6. What is the value of x?

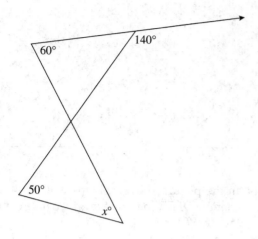

   A. 40
   B. 50
   C. 60
   D. 70

7. The base of a rectangle is three times as long as the height. If the perimeter is 64, what is the area of the rectangle?

   A. 24
   B. 64
   C. 96
   D. 192

8. The area of an equilateral triangle is $36\sqrt{3}$ cm². What is the height of the triangle?

   A. 6 cm
   B. 12 cm
   C. $6\sqrt{3}$ cm
   D. $12\sqrt{3}$ cm

9. $\triangle ABC$ is an equilateral triangle and $\triangle ADC$ is an isosceles triangle. If $AD = 12$ what is the area of the shaded region?

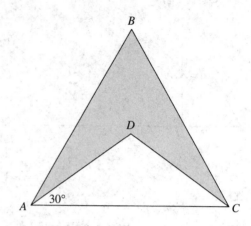

   A. $108\sqrt{23}$
   B. $72\sqrt{3}$
   C. $36\sqrt{3}$
   D. 144

10. Which of the following expressions represents the area of the shaded region below?

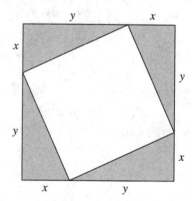

   A. $x^2 + y^2$
   B. $2(x^2 + y^2)$
   C. $2xy$
   D. $x^2 + 2xy + y^2$

EXPLAINED ANSWERS

1. **EXPLANATION: Choice D is correct.**

   The sum of the acute angles in a right triangle is 90°.

   $y + 20 = 90 \implies y = 70 \implies x + 40 = 90 \implies x = 50 \implies y - x = 70 - 50 = 20.$

2. **EXPLANATION: Choice A is correct.**

   The triangle is a 45°-45°-90° right triangle.

   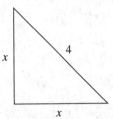

   $\sqrt{2}x = 4 \implies x = \dfrac{4}{\sqrt{2}} \times \dfrac{\sqrt{2}}{\sqrt{2}} = \dfrac{4\sqrt{2}}{\sqrt{2}} = 2\sqrt{2}.$

   The height and base measure $2\sqrt{2}$.
   Find the area.

   $A = \dfrac{1}{2} \times (2\sqrt{2}) \times (2\sqrt{2}) = \dfrac{1}{2} \times 4 \times \sqrt{2} \times \sqrt{2}$

   $\dfrac{1}{2} \times 4 \times 2 = 4 \text{ in}^2.$

3. **EXPLANATION: Choice B is correct.**

   The sum of the angles is 180°.

   $x + y + 50 = 180 \implies x + y = 130$ and $a + b + c = 180.$

   Therefore, $a + b + c + x + y = 180 + 130 = 310.$

4. **EXPLANATION: Choice C is correct.**

   The angles on the two triangles should be equal, even if the sides are not correct, because the triangles are similar. Therefore, no matter what the sides on the larger triangle are, the fraction representing tan E will be equivalent to the fraction representing tan B. Tangent means taking the opposite side over the adjacent side, so $\dfrac{4}{3}$.

5. **EXPLANATION: Choice A is correct.**

   The base and height for the triangle and rectangle are equal.
   The area of the rectangle is 120. The area of the triangle is half the area of the rectangle, $\dfrac{1}{2}(120) = 60.$

6. **EXPLANATION: Choice B is correct.**

    The sum of the angles is 180°. Use what you know about linear pairs and vertical angles to find the bold angle measures.

    $x + 50 + 80 = 180 \implies x + 130 = 180 \implies x = 50.$

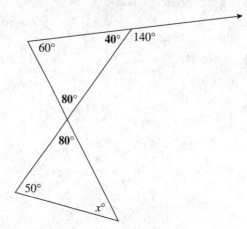

7. **EXPLANATION: Choice D is correct.**

    Draw a picture.

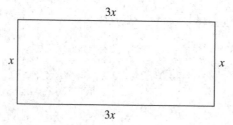

    The perimeter is 64 so $8x = 64 \implies x = 8$.

    So, $h = 8$ and $b = 3(8) = 24$.

    Substitute $h = 8$ and $b = 24$ in the area formula. $A = b \times h = 8 \times 24 = 192$.

8. **EXPLANATION: Choice C is correct.**

    Draw the height on the diagram to form two 30°-60°-90° triangles.

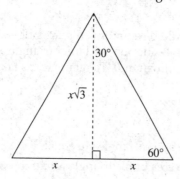

    Write the area formula in terms of x and simplify.

    $A = 36\sqrt{3} = \dfrac{1}{2} \times (2x) \times (x\sqrt{3})$

    Solve for x.

    $36\sqrt{3} = x^2\sqrt{3} \implies 36 = x^2 \implies 6 = x.$

    The height is $h = 6\sqrt{3}$.

9. **EXPLANATION: Choice B is correct.**

Draw the height of the unshaded triangle to form two 30°-60°-90° right triangles.

The height is across from the 30° angle, so it is half the length of the hypotenuse or 6.

The base is across from the 60° angle so the base measures $6\sqrt{3}$.

That means $AC = \left[2(6\sqrt{2})\right] = 12\sqrt{3}$.

Use the area formula.

Area $\triangle ADC = \dfrac{1}{2} \times 12\sqrt{3} \times 6 = 6\sqrt{3} \times 6 = 36\sqrt{3}$.

The area of the shaded region is twice the area of $\triangle ADC$ $2(36\sqrt{3}) = 72\sqrt{3}$.

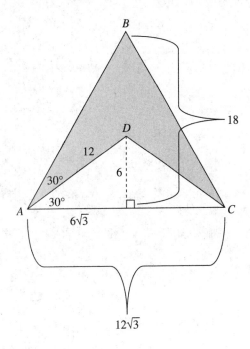

10. **EXPLANATION: Choice C is correct.**

The base of each of the shaded triangles is x, and the height of each is y. The area of each of the shaded triangles is

$A_T = \dfrac{1}{2}xy$. That means the area of the entire shaded region is $A_S = 4\left(\dfrac{1}{2}xy\right) = 2xy$.

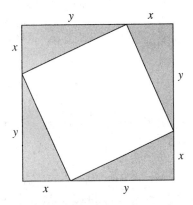

CHAPTER 24

CIRCLES

On the SAT you will be asked to solve problems that relate the area of a **circle** to the area of a **sector** and the **circumference** of a circle to the length of an **arc**. In addition, there will be questions that require knowledge of the **equation of a circle**. Begin with the mathematics review and then complete and correct the practice problems. There are 2 Solved SAT Problems and 11 Practice SAT Questions with answer explanations.

Circle—a set of points equidistant from a given point, the center.

Radius—the distance from the center of a circle to its perimeter.

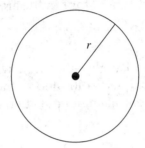

Chord: A chord is line segment with endpoints on a circle.

Diameter: The diameter is the distance through the center from one point on the circle to another.

The diameter is twice the length of the radius, $d = 2r$.

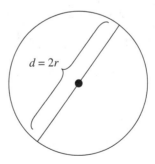

Circumference: The circumference is the perimeter of the circle, $C = 2\pi r = \pi d$.

The rotation around a circle is 360°, around a half of a circle is $\dfrac{360°}{2} = 180°$, and around a quarter of a circle is $\dfrac{360°}{4} = 90°$. A **radian** is another way to describe the amount of rotation around a circle. Rotating 360° is the same as rotating 2π *radians*. So 360° = 2π *radians*, 180° = π *radian* and 90° = $\dfrac{\pi}{2}$ *radians*.

Example 1.

What is the radius of a circle with circumference of 30π?

Substitute $C = 30\pi$ in the circumference formula.

$C = 2\pi r \Rightarrow 30\pi = 2\pi r \Rightarrow 15 = r$.

Area of a circle is $A = \pi r^2$.

Example 2.

What is the area of a circle with diameter 20?

Find the radius.

$d = 2r \Rightarrow 20 = 2r \Rightarrow 10 = r$.

Substitute 10 for r in the area formula.

$A = \pi r^2 \Rightarrow A = \pi \times 10^2 \Rightarrow A = 100\pi$.

Central angle—the angle with endpoints located on a circle's circumference and vertex located at the circle's center. $\angle AOC$ below is a central angle. Central angles partition a circumference into arcs.

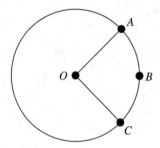

Arc—a piece of the circumference.

$$\frac{\text{Arc length}}{\text{Circumference}} = \frac{\text{Measure of central angle}}{360°}$$

Example 3.

What is the measure of arc ABC in circle O seen below?

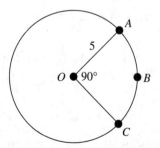

$\dfrac{90°}{360°} = \dfrac{1}{4}$. So the arc length is $\dfrac{1}{4}$ of the circumference $= \dfrac{1}{4}(1\pi) = \dfrac{5\pi}{2}$.

Sector—a piece of the area.

$$\frac{\text{Area of sector}}{\text{Area of circle}} = \frac{\text{Measure of central angle}}{360°}$$

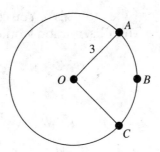

Example 4.

The area of a sector in a circle with radius 3 is 3π. What is the measure of the central angle?

Use 3π for the sector area and $\pi(3)^2 = 9\pi$ for the circle area $\dfrac{3\pi}{9\pi} = \dfrac{1}{3}$.

There are 360° in a circle so $\dfrac{1}{3}(360°) = 120°$, the measure of the central angle.

Tangent—a line that touches the circle at one point. In the figure below $\overleftrightarrow{AB}$ is tangent to circle O.

When a line is tangent to a circle, the line is perpendicular to the radius at the point of tangency. Therefore, in the figure below $m\angle OAB = 90°$.

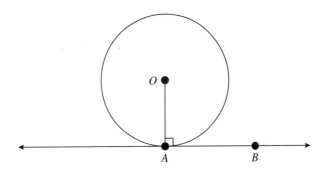

Example 5.

$\overleftrightarrow{AB}$ is tangent to circle O at point A. $OB = 13$, and $AB = 12$. What is the radius of the circle?

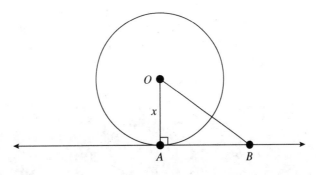

Use the Pythagorean Theorem. $(OA)^2 + (AB)^2 = (OB)^2$.

Substitute $AB = 12$ and $OB = 13$. $(OA)^2 + 12^2 = 13^2$.

Solve for OA. $(OA)^2 + 144 = 169 \Rightarrow (OA)^2 = 25 \Rightarrow OA = 5$.

Therefore, the radius is 5. You could also notice that the triangle is a 5-12-13 right triangle. This observation would allow you to find the length of the radius more quickly.

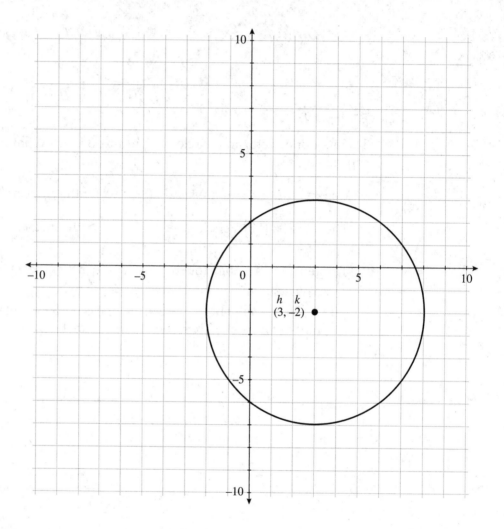

Standard Equation of a Circle

The standard **equation of a circle** is $(x - h)^2 + (y - k)^2 = r^2$. (h, k) is the center and r is the radius. The circle shown above has a center of $(3, -2)$ and a radius of 5. Therefore, the equation of the circle is $(x - 3)^2 + (y + 2)^2 = 5^2$. Write in standard form. $(x - 3)^2 + (y - (-2)^2 = 5^2$.

This equation is based on the Pythagorean Theorem.

Given below is the same circle with a right triangle drawn inside the figure. Consider the point on the circle (x, y).

The horizontal distance from the point (x, y) to the center $(3, -2)$ is $(x - 3)$.

The vertical distance from the point (x, y) to the center $(3, -2)$ is $(y - (-2))$.

These lengths along with the radius are drawn in the figure below. So, based on the Pythagorean Theorem, $a^2 + b^2 = c^2$, $a = (x - 3)$, $b = (y + 2)$, and $c = 5$ gives $(x - 3)^2 + (y + 2)^2 = 5^2$ $\rightarrow$ $(x - 3)^2 + (y - (-2))^2 = 25$. The standard form of the equation shows that when the center is $(3, -2)$ the radius is 5.

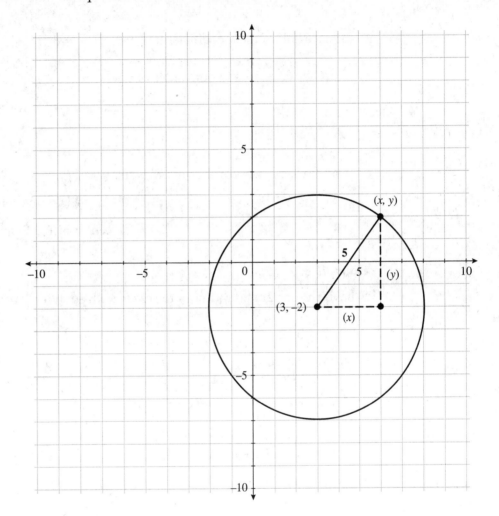

Example 6.

What is the center and radius of the circle with equation $(x + 4)^2 + (y - 5)^2 = 9$?

Write in standard form: $(x - (-4))^2 + (y - 5)^2 = 3^2$
The center is $(-4, 5)$ and the radius is $r\sqrt{9} = 3$.

Completing the Square

Sometimes the equation of the circle is not in standard from and is given in a more general form such as $x^2 + 12x + y^2 - 16y = -51$. Before answering any questions about the circle, rewrite the equation in standard form $(x - h)^2 + (y - k)^2 = r^2$, where (h, k) is the center and r is the radius.

Example 7.

What is the center and radius of the circle with equation $x^2 - 12x + y^2 + 16y = -51$?

Group variables: $\qquad\qquad\qquad\qquad x^2 - 12x \quad +y^2 + 16y \quad = -51$

Square half of the x-coefficient: $\qquad\qquad \dfrac{-12}{2} = -6. \quad \rightarrow \quad (-6)^2 = 36$

Square half of the y-coefficient: $\qquad\qquad \dfrac{16}{2} = 8 \quad \rightarrow \quad (8)^2 = 64$

Add above results to both sides of equation:

$$x^2 - 12x + \mathbf{36} + y^2 + 16y + \mathbf{64} = -51 + \mathbf{36} + \mathbf{64}$$

Simplify $\qquad\qquad\qquad\qquad \underbrace{x^2 - 12x + \mathbf{36}} + \underbrace{y^2 + 16y + \mathbf{64}} = \mathbf{49}$

Factor the perfect squares in brackets: $\qquad (x - 6)^2 \quad + \quad (y + 8)^2 = 7^2$

$$\Rightarrow (x - 6)^2 \quad + \quad (y - (-8))^2 = 7^2$$

The center is $(6, -8)$ and the radius is $r = 7$.

Example 8.

What is the center and radius of the circle with equation $2x^2 + 8x + 2y^2 - 4y = 40$?

Divide by 2: $\qquad\qquad\qquad\qquad x^2 + 4x + y^2 - 2y = 20$

Group variables: $\qquad\qquad\qquad\quad x^2 + 4x \quad +y^2 - 2y \quad = 20$

Square half of the x-coefficient: $\qquad\qquad \dfrac{4}{2} = 2. \quad \rightarrow \quad (2)^2 = 4$

Square half of the y-coefficient: $\qquad\qquad \dfrac{-2}{2} = -1 \quad \rightarrow \quad (-1)^2 = 1$

Add above results to both sides of the equation:

$$x^2 + 4x + \mathbf{4} + y^2 - 2y + \mathbf{1} = 20 + \mathbf{4} + \mathbf{1}$$

Simplify: $\qquad\qquad\qquad\qquad \underbrace{x^2 + 4x + \mathbf{4}} + \underbrace{y^2 - 2y + \mathbf{1}} = 25$

Factor the perfect squares in brackets: $\qquad (x + 2)^2 \quad + \quad (y - 1)^2 = 25$

$$\Rightarrow (x - (-2))^2 + (y - 1)^2 = 5^2$$

The center is $(-2, 1)$ and the radius is $r = 5$.

Practice Questions

1. What is the diameter of a circle with an area 121π?

2. The length of an arc in a circle with circumference 14π is 7π. What is the measure of the central angle?

3. What is the area of the shaded sector in circle O seen below?

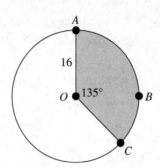

4. $\overleftrightarrow{AB}$ is tangent to circle O at point A. The area of $\triangle OAB$ is 224. What is the value of AB?

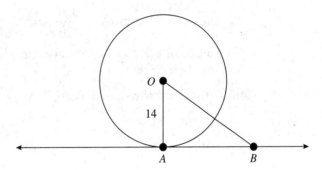

5. What are the radius and center of the circle with equation
   $2x^2 - 20x + 2y^2 + 12y = 6$?

Practice Answers

1. 22.
2. 180°.
3. 96π.
4. 32.
5. Center is (5, –3); radius is square root of 37.

■■■ **SOLVED SAT PROBLEMS**

---

1. Point A is the center of the circle below. What is the radius of the circle if the area of the shaded region is 27π?

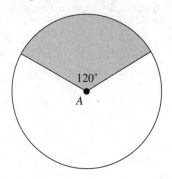

EXPLANATION: The correct answer is 9.

Since there are $120°$ in the shaded sector and $360°$ in the in the total circle, the area of the circle is $\dfrac{360°}{120°} = 3$ times the area of the shaded region.

Therefore, the area of the shaded region is $3(27\pi) = 81\pi$. The formula for area of a circle is $A = \pi r^2$.

Substitute 81π for A: $81\pi = \pi r^2$

Divide by π: $81 = r^2$

Take the square root of both sides: $9 = r$.

---

2. Which of the following is an equation of a circle in the xy-plane with center $(-5, 4)$ and point $(3, -2)$ on the circle?

   A. $(x - 5)^2 + (y + 4)^2 = 10$
   B. $(x + 5)^2 + (y - 4)^2 = 10$
   C. $(x - 5)^2 + (y + 4)^2 = 100$
   D. $(x + 5)^2 + (y - 4)^2 = 100$

EXPLANATION: Choice D is correct.

$(-5, 4)$ is the center, producing an equation $(x + 5)^2 + (y - 4)^2 = r^2$.

$(3, -2)$ is on the circle.

Substitute $x = 3$ and $y = -2$ into above equation: $(3 + 5)^2 + (-2 - 4)^2 = r^2$

Simplify: $(8)^2 + (-6)^2 = r^2$ $\rightarrow$ $64 + 36 = r^2$ $\rightarrow$ $100 = r^2$

To find r, take the square root on both sides, which gives $10 = r$, but in reality, to find equation, we just need to know that $100 = r^2$.

So, the equation of circle is $(x + 5)^2 + (y - 4)^2 = 100$.

CIRCLES
PRACTICE SAT QUESTIONS

ANSWER SHEET

Choose the correct answer.
If no choices are given, grid the answers in the section at the bottom of the page.

1. (A) (B) (C) (D)   11. (A) (B) (C) (D)
2. (A) (B) (C) (D)
3. (A) (B) (C) (D)
4. **GRID**
5. (A) (B) (C) (D)
6. (A) (B) (C) (D)
7. (A) (B) (C) (D)
8. (A) (B) (C) (D)
9. (A) (B) (C) (D)
10. (A) (B) (C) (D)

Use the answer spaces in the grid below if the question requires a grid-in response.

| Student-Produced Responses | ONLY ANSWERS ENTERED IN THE CIRCLES IN EACH GRID WILL BE SCORED. YOU WILL NOT RECEIVE CREDIT FOR ANYTHING WRITTEN IN THE BOXES ABOVE THE CIRCLES. |

4.

PRACTICE SAT QUESTIONS

1. What is the diameter of a circle whose area is 4?

   A. $\dfrac{2}{\sqrt{\pi}}$

   B. $\dfrac{4}{\sqrt{\pi}}$

   C. $4\sqrt{\pi}$

   D. 2π

2. What is the area of the shaded region?

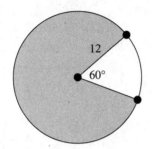

   A. 24π
   B. 60π
   C. 90π
   D. 120π

3. A circle has the equation $(x - 3)^2 + (y + 7)^2 = 25$. What is the diameter?

   A. 3
   B. 5
   C. 10
   D. 25

4.

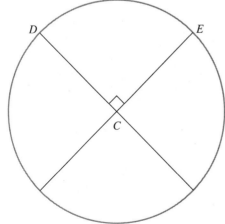

   The circle seen above has a center at C and an area of 16π. If $x\pi$ is the length of minor arc $\overset{\frown}{DE}$, find x.

5. The inner circle is tangent to the outer circle at point P. Point O is the center of the outer circle and $\overline{PO}$ is the diameter of the inner circle. What is the ratio of the area of the outer circle to the area of the inner circle?

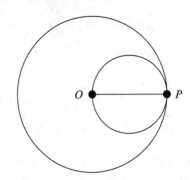

   A. $2:1$
   B. $3:1$
   C. $3:2$
   D. $4:1$

6. The area of the circle O below is 81π. What is the length of $\overline{BA}$?

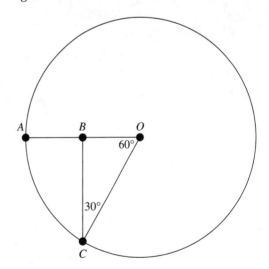

   A. 4.5
   B. 5
   C. 5.5
   D. 6

7. The area of the △*ABC* is 18, what is the area of the shaded region?

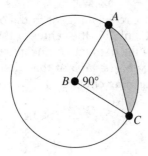

A. 18
B. 9π
C. 9π − 18
D. 9π + 18

8. The area of a circle is doubled. What is the ratio of the circle's new radius to the circle's original radius?

A. $\dfrac{\sqrt{2}}{1}$

B. $\dfrac{\sqrt{3}}{1}$

C. $\dfrac{4}{1}$

D. $\dfrac{1}{2}$

9. What is the value of *x* in circle *O* seen below?

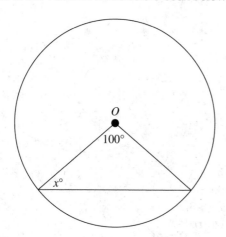

A. 20
B. 30
C. 40
D. 50

10. Bill draws a circle with the equation $(x - 4)^2 + (y + 2)^2 = 9$. All of these points are on or within the circle EXCEPT:

A. (2, 4)
B. (5, 0)
C. (4, −2)
D. (4, −5)

11. The diameter of the large circle is 12. The center of each of the three congruent smaller circles lies on the diameter of the larger circle. What is the area of the shaded region?

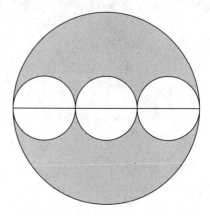

A. 12π
B. 18π
C. 24π
D. 36π

EXPLAINED ANSWERS

1. **EXPLANATION: Choice B is correct.**

   Substitute $A = 4$ in the area formula and solve for r.

   $$A = 4 = \pi r^2 \Rightarrow \frac{4}{\pi} = r^2 \Rightarrow \sqrt{\frac{4}{\pi}} = r \Rightarrow \frac{2}{\sqrt{\pi}} = r$$

   The diameter is twice the radius. $d = 2\frac{2}{\sqrt{\pi}} = \frac{4}{\sqrt{\pi}}$

2. **EXPLANATION: Choice D is correct.**

   The circle is 360° and the shaded region is 300°. Write a fraction.

   $\dfrac{\text{Area of Shaded Region}}{\text{Area of Circle}} = \dfrac{300}{360} = \dfrac{5}{6}$. The shaded region is $\dfrac{5}{6}$ of the circle.

   Area of circle $= \pi r^2 = \pi(12)^2 = 144\pi$

   $144\pi \times \dfrac{5}{6} = 120\pi$.

3. **EXPLANATION: Choice C is correct.**

   The equation of a circle has the radius squared as the right side of the equation: thus, the radius is 5. If the radius is 5, then the diameter is 10.

4. **EXPLANATION: The correct answer is 2.**

   All parts of a circle are proportional, so the minor arc is $\dfrac{1}{4}$ of the circumference since the sector of area it corresponds to is $\dfrac{1}{4}$ of the area. The radius of the circle is 4 (work backwards from the area!), so the diameter is 8 and the circumference is 8π. Divide by 4 to get 2π, making $x = 2$.

5. **EXPLANATION: Choice D is correct.**

   Let d be the diameter of the inner circle. That means the <u>radius</u> of the outer circle is d, and the radius of the inner circle is $\dfrac{d}{2}$. The area of the outer circle is πd^2.

   The area of the inner circle is $= \pi\left(\dfrac{d}{2}\right)^2 = \dfrac{\pi d^2}{4}$

   The ratio of the outer circle to the inner circle is $\dfrac{\pi d^2}{\dfrac{\pi d^2}{4}} = 4 : 1$.

6. **EXPLANATION: Choice A is correct.**

   Substitute 81π for A. $A = 81\pi = \pi r^2$

   Solve for r. $81 = r^2 \Rightarrow 9 = r$.

   That means $OC = 9$ and $OA = 9$.

   $OB = 4.5$ and $OA = 9$ so $BA = 9 - 4.5 = 4.5$.

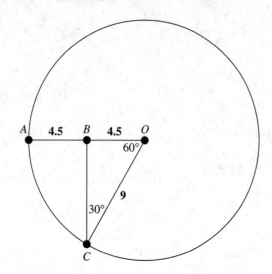

7. **EXPLANATION: Choice C is correct.**

   Area of $\triangle ABC = 18 = \dfrac{1}{2} \times b \times h$. The base and the height of the triangle are both radii.

   Substitute r for b and h in the area formula.

   Area of $\triangle ABC = 18 = \dfrac{1}{2} \times r^2$

   Solve for r. $36 = r^2 \Rightarrow 6 = r$.

   Area of circle is $\pi(6)^2 = 36\pi$

   The central angle is $90°$ and area of the circle is 36π.

   Area of sector $= \dfrac{90°}{360°}(36\pi) = \dfrac{1}{4}(36\pi) = 9\pi$

   The area of the shaded region is area of sector − area of triangle $= 9\pi - 18$.

8. **EXPLANATION: Choice A is correct.**

   A_1 is old area. A_2 is new area.

   $2A_1 = A_2$.

   $2\pi r_1^2 = \pi r_2^2$. Cancel pi

   $2r_1^2 = r_2^2$.

   We're looking for ratio of r_2 to r_1. So divide both sides by r_1^2.

   Now:

   $2 = \dfrac{r_2^2}{r_1^2}$. Take the square root of both sides.

   $\dfrac{r_2}{r_1} =$ square root of 2.

9. **EXPLANATION: Choice C is correct.**

   The triangle has two radii as sides. The angles across from these sides must be equal.

   The sum of the angles in a triangle is 180°, so $2x + 100 = 180$.

   Solve for x. $2x = 80 \Rightarrow x = 40$.

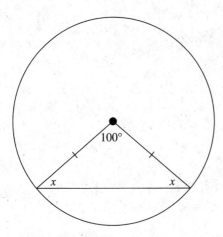

10. **EXPLANATION: Choice A is correct.**

    The center of Bill's circle, according to the equation, is $(4, -2)$ and the radius is 3. You can solve this by plugging in the points one by one to the given equation (note that the left side would be less than or equal to 9 for all points within the circle), but it may be faster and easier to just draw a rough picture: $(2, 4)$ is too high on the y-axis to be part of the circle.

11. **EXPLANATION: Choice C is correct.**

    The diameter of the large circle is 12, so the radius is 6.

    The area of the large circle is $\pi \cdot 6^2 = 36\pi$.

    The radius of each small circle is $\dfrac{6}{3} = 2$. The area of each small circle is $\pi \cdot 2^2 = 4\pi$.

    The area of the shaded region is $36\pi - 3(4\pi) = 36\pi - 12\pi = 24\pi$.

CHAPTER 25

VOLUME

On the SAT you will be asked to find the **volume** of three-dimensional figures: cylinder, rectangular prism, cone, and sphere. There will also be questions where the volume of the figure is given and you are asked to compute the measurement of some part of the figure. Begin with the mathematics review and then complete and correct the practice problems. There are 2 Solved SAT Problems and 10 Practice SAT Questions with answer explanations.

Example 1.

Find the volume of a right circular cylinder that has a diameter of 10 inches and a height of 20 inches. Round your answer to the nearest whole number.

The formula for volume of a cylinder is $V = \pi r^2 h$*.

Since the diameter is 10 in, the radius is 5 in.

Substitute $r = 5$ and $h = 20$ into the volume formula: $V = \pi(5)^2(20)$

Solve for V: $V = \pi(5)^2(20) = \pi(25)(20) = 500\pi \approx 1{,}570$ if you use 3.14 for pi or 1571 if you use the pi key on your calculator. $V \approx 1{,}570 \text{ in}^3$ or $1{,}571 \text{ in}^3$

Example 2.

What is the radius of a right circular cylinder that has a height of 12 cm and a volume of $300\pi \text{cm}^3$?

The formula for volume of a cylinder is $V = \pi r^2 h$.

Substitute $V = 300\pi$ and $h = 12$ into the volume formula: $300\pi = \pi r^2(12)$

Divide both sides of equation by 12π: $25 = r^2$

Take the square root of both sides of the equation: $5 = r \quad \rightarrow \quad r = 5 \text{ cm}$

Example 3.

Find the volume of a right circular cone that has a radius of 7 in and a height of 15 in.

The formula for volume of a cone is $V = \dfrac{\pi r^2 h}{3}$.

Substitute $r = 7$ and $h = 15$ into the volume formula: $V = \dfrac{\pi(7)^2(15)}{3}$

Solve for V: $V = \dfrac{\pi(7)^2(15)}{3} = \dfrac{\pi(49)(15)}{3} = \pi(49)(5) = 245\pi \text{ in}^3$

Example 4.

What is the diameter of a right circular cone that has a height of 10 cm and a volume of $270\pi \text{ cm}^3$?

The formula for volume of a cone is $V = \dfrac{\pi r^2 h}{3}$.

Substitute $V = 270\pi$ and $h = 10$ into the volume formula: $270\pi = \dfrac{\pi r^2 10}{3}$

---

*Volume formulas are at the beginning of the section.

Multiply both sides by 3: $810\pi = \pi r^2 10$

Divide both sides of equation by 10π: $81 = r^2$

Take the square root of both sides of the equation: $9 = r \rightarrow d = 18$ cm.

Example 5.

Find the height of a rectangular prism if the volume is 4,862 in³, the width is 17 in, and the length is 13 in.

The formula for volume of a rectangular prism is $V = lwh$.

Substitute $V = 4{,}862$, $w = 17$, and $l = 13$ into the volume formula: $4{,}862 = (13)(17)h$

Simplify the right side of equation: $4{,}862 = 221h$

Divide both sides of equation by 221: $22 = h \rightarrow h = 22$ in.

Example 6.

Find the volume of a sphere that has a diameter of 18 cm. Round your answer to the nearest whole number.

The formula for volume of a sphere is $V = \dfrac{4\pi r^3}{3}$.

Since the diameter is 18 cm the radius is 9 cm.

Substitute $r = 9$ into the volume formula: $V = \dfrac{4\pi(9)^3}{3}$

Solve for V: $V = \dfrac{4\pi(729)}{3} = 4\pi(243) = 972\pi \approx 3052$ if you use 3.14 for pi or 3054 if you use the pi key on your calculator. $\rightarrow V \approx 3052$ cm³ or 3,054 cm³.

Example 7.

What is the radius of a sphere that has a volume of is 288π in³?

The formula for volume of a sphere is $V = \dfrac{4\pi r^3}{3}$.

Substitute $V = 288\pi$ into the volume formula: $288\pi = \dfrac{4\pi r^3}{3}$

Multiply both sides by 3: $864\pi = 4\pi r^3$

Divide both sides of the equation by 4π: $216 = r^3$

Take the cube root of both sides of the equation: $6 = r \rightarrow r = 6$ in.

Practice Questions

1. What is the diameter of a right circular cylinder that has a height of 14 cm and a volume of 126π cm³?
2. Find the volume of a right circular cone that has a radius of 5 in and a height of 18 in. Round your answer to the nearest whole number.
3. What is the radius of a right circular cone that has a height of 9 cm and a volume of 147π cm³?
4. Find the length of a rectangular prism if the volume is 1,890 in³, the width is 15, and the height is 7 in.
5. Find the volume of a sphere that has a diameter of 24 cm. Round your answer to the nearest whole number.
6. What is the diameter of a sphere that has a volume of 972π in³?

Practice Answers

1. $d = 6$ cm.
2. $V = 150\pi$ in^3 or approximately 471.
3. $r = 7$ cm.
4. $l = 18$ in.
5. $V = 2,304\pi$ cm^3 or approximately 7235 cm^3 if you use 3.14 for pi or 7238 cm^3 if you use the pi key on your calculator.
6. $d = 18$ in.

SOLVED SAT PROBLEMS

1. What is the diameter of a cylindrical glass with a height of 10 cm and a volume of 160π cm^3?

   EXPLANATION: The correct answer is 8.

   The formula for volume of a cylinder is $V = \pi r^2 h$.

   Substitute $V = 160\pi$ and $h = 10$ into the volume formula: $160\pi = \pi r^2(10)$

   Divide both sides of the equation by 10π: $16 = r^2$.

   Take the square root of both sides of the equation: $4 = r \rightarrow d = 8$ cm.

2. A cylindrical pool has a diameter of 16 feet and a height of 5 feet. If the pool is completely empty and it takes 5 hours at a constant flow rate to completely fill the pool, which of the following is closest to the flow rate, in gallons per minute? (1 cubic foot is equivalent to 7.48 gallons.)

   A. 15

   B. 20

   C. 25

   D. 30

   EXPLANATION: Choice C is correct.

   First find the volume of the pool. The formula for the volume of a cylinder is $V = \pi r^2 h$. Since the diameter is 16 feet, the radius is 8 feet.

   Substitute $r = 8$ and $h = 5$ into the volume formula: Using the pi key. $V = \pi(8)^2(5) = 320\pi \approx 1,005.31$ ft^3

   Since this volume is in cubic feet, convert into gallons by multiplying by 7.48.

   $V \approx 1,005.31$ ft^3 $\rightarrow$ $V \approx 1,005.31(7.48) \approx 7,519.72$ gallons

   Divide 7,519.72 by 5 to convert in gallons per hour:
   $$\frac{7,519.72}{5} \approx 1,503.94 \text{ gallons per hour}$$
   Divide 1,503.944 by 60 to convert in gallons per minute :
   $$\frac{1,503.94}{60} \approx 25.07 \text{ gallons per minute. This result is based on using the}$$
   pi key. The correct answer choice is the same if you used 3.14 for pi.

   Choice C, 25, is the closest to the flow rate in gallons per minute.

VOLUME
PRACTICE SAT QUESTIONS

ANSWER SHEET

Choose the correct answer.
If no choices are given, grid the answers in the section at the bottom of the page.

1. (A) (B) (C) (D)
2. (A) (B) (C) (D)
3. (A) (B) (C) (D)
4. GRID
5. (A) (B) (C) (D)
6. GRID
7. (A) (B) (C) (D)
8. (A) (B) (C) (D)
9. (A) (B) (C) (D)
10. (A) (B) (C) (D)

Use the answer spaces in the grids below if the question requires a grid-in response.

Student-Produced Responses   ONLY ANSWERS ENTERED IN THE CIRCLES IN EACH GRID WILL BE SCORED. YOU WILL NOT RECEIVE CREDIT FOR ANYTHING WRITTEN IN THE BOXES ABOVE THE CIRCLES.

4.

6.

PRACTICE SAT QUESTIONS

1. What is the height of a right cylinder with radius 5 in and volume 150π in^3?

    A. 5 in
    B. 6 in
    C. 7 in
    D. 8 in

2. What is the surface area of a cube with a volume of 5,832 ft^3?

    A. 2,916 ft^2
    B. 2,430 ft^2
    C. 1,944 ft^2
    D. 1,458 ft^2

3. A sphere is created with half the radius of the original sphere. What is the ratio of the volume of the original sphere to the volume of the new sphere?

    A. 1 : 8
    B. 8 : 1
    C. 2 : 1
    D. 1 : 2

4. The volume of a rectangular prism with a square base is 128 cubic inches. If the height of the prism is 8, what is the perimeter of the base?

5. The area of the base of a cylinder is 100π m^2. The volume of the cylinder is 900π m^2. What is the height of the cylinder?

    A. 9 m
    B. 10 m
    C. 11 m
    D. 12 m

6. A canister of instant coffee is filled with a mix of decaffeinated and regular coffee to exactly 2 cm from the top before it is labeled and sent to be shipped. The canister's base has an area of 50 cm^2, and the height is 12 cm. If 200 cm^3 of decaffeinated coffee is in the canister, what is the volume of the regular coffee that must be added?

7. What is the volume of a regular hexagonal prism with height 13 ft and base-edge length of 10 ft?

    A. $1,950\sqrt{3}$
    B. $750\sqrt{3}$
    C. $150\sqrt{3}$
    D. $25\sqrt{3}$

8. What is the radius of a right cone with height 15 cm and volume 90π cm^3?

    A. $3\sqrt{2}$ cm
    B. $2\sqrt{3}$ cm
    C. 6 cm
    D. 9 cm

9. The conical tank below is filled with liquid. How much of the tank is empty if the water level is 9 feet from the tip?

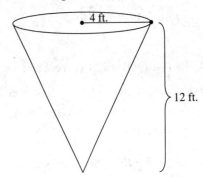

    A. 67π ft^3
    B. 57π ft^3
    C. 47π ft^3
    D. 37π ft^3

10. The figure below is the side view of a pool that is 10 feet wide. What is the volume of the water in the pool when it is filled to 2 feet below the top?

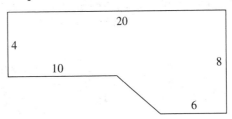

    A. 1,120
    B. 720
    C. 360
    D. 200

EXPLAINED ANSWERS

1. **EXPLANATION: Choice B is correct.**

    Substitute $V = 150\pi$ and $r = 5$.
    $V = \pi r^2 h = 150\pi \Rightarrow \pi(5)^2 h = 150\pi$
    Solve for h. $25\pi h = 150\pi \Rightarrow h = 6$ in.

2. **EXPLANATION: Choice C is correct.**

    Substitute $V = 5832$. $V = s^3 = 5,832$.
    Solve for s. $s = 18$.
    Substitute $s = 18$. $SA = 6\,s^2 = 6(18)^2 = 1,944$ ft².

3. **EXPLANATION: Choice B is correct.**

    The volume of the original sphere is $\dfrac{4}{3}\pi r^3$.

    The volume of the new sphere is $\dfrac{4}{3}\pi\left(\dfrac{r}{2}\right)^3 = \dfrac{4}{3}\pi\dfrac{r^3}{8}$.

    $\dfrac{\text{Volume of original sphere}}{\text{Volume of new sphere}} = \dfrac{\dfrac{4}{3}\pi r^3}{\dfrac{4}{3}\pi\dfrac{r^3}{8}} = \dfrac{1}{\frac{1}{8}} = \dfrac{8}{1} = 8:1$.

4. **EXPLANATION: The correct answer is 16.**

    Divide volume by height to find area of the square base. $128 \div 8 = 16$.
    Take square root of area to find length of each side. $\sqrt{16} = 4$.
    Multiply length of side by 4 to find perimeter of square. $4 \cdot 4 = 16$.

5. **EXPLANATION: Choice A is correct.**

    Find the area of the base.
    $A = \pi r^2 = 100\pi$
    Substitute $\pi r^2 = 100\pi$ and $V = 900\pi$.
    $V = \pi r^2 h \Rightarrow 900\pi = 100\pi h \Rightarrow 9\text{ m} = h$

6. **EXPLANATION: The correct answer is 300.**

    The volume of the canister is 50×12, but it isn't filled all the way, so subtract 2 cm from the height as the problem specifies. $50 \times 10 = 500$. 200 of that 500 volume is already taken up by decaffeinated coffee, leaving 300 cm³ of space for the regular coffee.

7. **EXPLANATION: Choice A is correct.**

First find the area of the hexagonal base. We know the length of each side of the hexagon is 10. We can find the area of one of the central equilateral triangles and multiply that area by 6. Use the rules of a 30°-60°-90° right triangle to find the height of the triangle is $5\sqrt{3}$.

The area of the triangle is

$$A = \frac{1}{2} \times 10 \times 5\sqrt{3} = 25\sqrt{3}$$

Area of hexagon is $6(25\sqrt{3}) = 150\sqrt{3}$ ft².

Substitute $B = 150\sqrt{3}$ and $h = 3$. The volume of the prism is $V = Bh = (150\sqrt{3}) \times 13 = 1{,}950\sqrt{3}$ ft³.

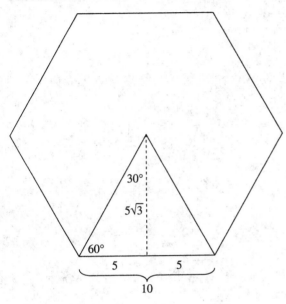

8. **EXPLANATION: Choice A is correct.**

Substitute $V = 90\pi$ and $h = 15$.

$$V = \frac{1}{3}\pi r^2 h \Rightarrow \frac{1}{3}\pi r^2(15) = 90\pi \Rightarrow 5\pi r^2 = 90\pi$$

Solve for r. $r^2 = 18 \Rightarrow r = \sqrt{18} \Rightarrow r = \sqrt{9 \cdot 2} = 3\sqrt{2}$

9. **EXPLANATION: Choice D is correct.**

Use the cone volume formula to find the original amount of liquid in the cone.

Substitute $r = 4$ and $h = 12$ to solve for V.

$$V = \frac{1}{3}\pi r^2 h = \frac{1}{3}\pi(4)^2 12 = \frac{192\pi}{3} = 64\pi \text{ ft}^3.$$

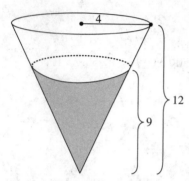

The solution finds the volume of the shaded portion. Then we subtract the volume of the shaded portion from the entire volume to find the volume of the unshaded portion. The volume of the unshaded portion is the correct answer.

When the liquid level decreases to 9 feet, the remaining liquid forms a new cone whose height is 9. We must now find the radius of this new cone. The ratio of radius to height stays constant.

Substitute $h = 9$. $\dfrac{r}{h} = \dfrac{4}{12}\dfrac{r}{9} = \dfrac{1}{3} \Rightarrow 3r = 9 \Rightarrow r = 3$.

Substitute $r = 3$ and $h = 9$.

Solve for r. The new amount of liquid is

$$V = \frac{1}{3}\pi(3)^2 9 = \frac{81\pi}{3} = 27\pi \text{ ft}^3.$$

So the amount of liquid that has poured out is the difference between the two volumes. $64\pi \text{ ft}^3 - 27\pi \text{ ft}^3 = 37\pi \text{ ft}^3$.

10. **EXPLANATION: Choice B is correct.**

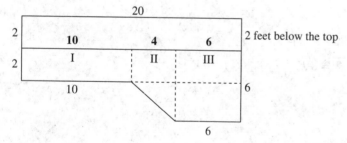

Partition the diagram into regions.

Region I (Rectangle)—$A_I = bh = 2 \cdot 10 = 20$

Region II (Trapezoid)—$A_{II} = \dfrac{1}{2} \cdot h(b_1 + b_2) = \dfrac{1}{2} \cdot 4(2 + 6) = \dfrac{1}{2} \cdot 4 \cdot 8 = 16$

Region III (Square)—$A_{III} = bh = 6 \cdot 6 = 36$

The area of the figure is $20 + 16 + 36 = 72$.

The volume is width times area: $10 \cdot 72 = 720$.

CHAPTER 26

TRIGONOMETRY

The trigonometry questions on the SAT draw on your knowledge of **sine** (*sin*), **cosine** (*cos*), and **tangent** (*tan*). That includes knowing how the sine and cosine of complementary angles are related, along with the relationship the tangents of complementary angles. In addition, you will be asked questions about how the sine, cosine, and tangent of angles of similar triangles are related. Begin with the mathematics review and then complete and correct the practice problems. There are 2 Solved SAT Problems and 11 Practice SAT Questions with answer explanations.

The acronym SOHCAHTOA is often used to help remember the rules for sine, cosine, and tangent.

$$\sin(\text{angle}) = \frac{\text{opposite}}{\text{hypotenuse}} = \frac{O}{H} \qquad \cos(\text{angle}) = \frac{\text{adjacent}}{\text{hypotenuse}} = \frac{A}{H}$$

$$\tan(\text{angle}) = \frac{\text{opposite}}{\text{adjacent}} = \frac{O}{A}$$

Example 1.

In the triangle below, find $\sin(\angle C)$, $\cos(\angle C)$, and $\tan(\angle C)$.

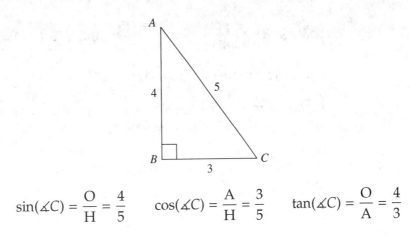

$$\sin(\angle C) = \frac{O}{H} = \frac{4}{5} \qquad \cos(\angle C) = \frac{A}{H} = \frac{3}{5} \qquad \tan(\angle C) = \frac{O}{A} = \frac{4}{3}$$

Example 2.

In the triangle below, find $\sin(\angle D)$ and $\cos(\angle F)$. What is the relationship between these values?

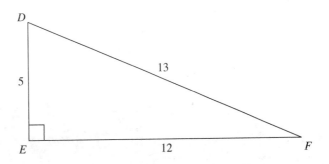

$$\sin(\angle D) = \frac{\text{opposite}}{\text{hypotenuse}} = \frac{O}{H} = \frac{12}{13} \quad \text{and} \quad \cos(\angle F) = \frac{\text{adjacent}}{\text{hypotenuse}} = \frac{A}{H} = \frac{12}{13}.$$

$\sin(\angle D)$ and $\cos(\angle F)$ are equal.

In this example, the opposite of one angle is the adjacent of the other. This is why $\sin(\angle D)$ and $\cos(\angle F)$ are equal.

Notice that $\angle D$ and $\angle F$ are complementary because the sum of $m\angle D$ and $m\angle F$ is 90°. From this we have a general rule that the sine and cosine of complementary angles are equal.

Example 3.

In the triangle below, find $\tan(\angle X)$ and $\tan(\angle Z)$. What is the relationship between these values?

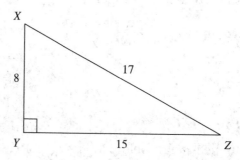

So $\tan(\angle X) = \frac{\text{opposite}}{\text{adjacent}} = \frac{O}{A} = \frac{15}{8} \quad \text{and} \quad \tan(\angle Z) = \frac{\text{opposite}}{\text{adjacent}} = \frac{O}{A} = \frac{8}{15}.$

$\tan(\angle X)$ and $\tan(\angle Z)$ are reciprocals.

In this example, the opposite of one angle is the adjacent of the other. This is why $\tan(\angle X)$ and $\tan(\angle Z)$ are reciprocals.

Notice that $\angle X$ and $\angle Z$ are complementary because the sum of $m\angle D$ and $m\angle F$ is 90°. From this we have a general rule that the tangent of complementary angles are reciprocals.

Example 4.

In the diagram below, ΔJKL is similar to ΔPQR, $\Delta JKL\sim\Delta PQR$. Find $\tan(\angle P)$.

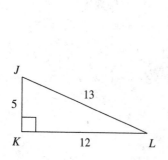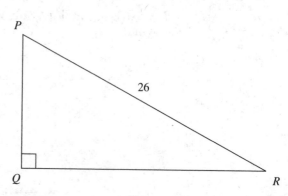

Since the triangles are similar, the tangents of corresponding angles are equal. The way to solve this problem is to use $\triangle JKL$ for which the values are know, and notice that $\tan(\angle P) = \tan(\angle J) = \dfrac{12}{5}$. From this we have a general rule that the tangent of corresponding angles of similar triangles are equal. It is also a general rule that the sine and cosine of corresponding angles of similar triangles are equal.

Practice Questions

1. In the triangle shown below, if $\sin(\angle C) = \dfrac{3}{10}$, then $\cos(\angle A) = ?$

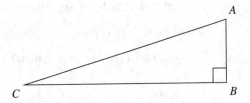

2. In $\triangle DEF$, $m\angle E = 90°$. If $\tan(\angle F) = \dfrac{5}{7}$, then $\tan(\angle D) = ?$

3. In $\triangle JKL$, $m\angle K = 90°$, $JK = 9$, $KL = 12$, and $LJ = 15$. $\triangle PQR$ is similar to $\triangle JKL$ and each side in $\triangle PQR$ is 5 times the length of the corresponding sides in $\triangle JKL$. What is the value of $\sin(\angle P)$?

Practice Answers

1. $\cos(\angle A) = \dfrac{3}{10}$.

2. $\tan(\angle D) = \dfrac{7}{5}$.

3. $\sin(\angle P) = \dfrac{12}{15} = \dfrac{4}{5}$.

SOLVED SAT PROBLEMS

1. $\angle X$ and $\angle Y$ are acute angles, $\sin(\angle X) = \cos(\angle Y)$, and the measure of $\angle X$ is 30 less than twice the measure of $\angle Y$. Find the measure of $\angle Y$.

   EXPLANATION: The correct answer is 40.

   Since $\sin(\angle X) = \cos(\angle Y)$, $x + y = 90$, and since the measure of $\angle X$ is 30 less than twice the measure of $\angle Y$, $x = 2y - 30$. Therefore, we have two equations: $x + y = 90$ and $x = 2y - 30$.

   Substitute $x = 2y - 30$ into $x + y = 90$: $(2y - 30) + y = 90$

   Simplify left side of equation: $3y - 30 = 90$

   Add 30 to both sides of equation: $3y = 120$

   Divide both sides of equation by 3: $y = 40$

2.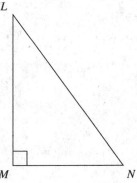

In the figure above, $\cos(\angle L) = \dfrac{4}{5}$. If $LM = 20$, what is the length of $\overline{MN}$?

A. 3

B. 5

C. 15

D. 25

EXPLANATION: Choice C is correct.

The $\cos(\angle L) = \dfrac{A}{H} = \dfrac{4}{5} = \dfrac{20}{H}$

Cross Multiply: $4H = 100$. $H = 25$.

Use the Pythagorean Theorem to solve for MN:

Set up Pythagorean Theorem: $(MN)^2 + (20)^2 = (25)^2$

Square numbers: $(MN)^2 + 400 = 625$

Subtract 400 from both sides of equation: $(MN)^2 = 225$

Take square root of both sides: $MN = 15$

You might have noticed this to be a 3-4-5 right triangle to find that $MN = 15$.

TRIGONOMETRY
PRACTICE SAT QUESTIONS

Choose the correct answer.
If no choices are given, grid the answers in the section at the bottom of the page.

1. GRID

2. GRID

3. (A) (B) (C) (D)

4. GRID

5. (A) (B) (C) (D)

6. (A) (B) (C) (D)

7. GRID

8. GRID

9. (A) (B) (C) (D)

10. (A) (B) (C) (D)

11. GRID

Use the answer spaces in the grids below if the question requires a grid-in response.

Student-Produced Responses ONLY ANSWERS ENTERED IN THE CIRCLES IN EACH GRID WILL BE SCORED. YOU WILL NOT RECEIVE CREDIT FOR ANYTHING WRITTEN IN THE BOXES ABOVE THE CIRCLES.

1.

2.

4.

7.

8.

11.

PRACTICE SAT QUESTIONS

1. In the right triangle shown below, sin(∠BAC) = 0.3. What is the value of cos(∠BCA)?

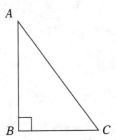

2. In the right triangle shown below, cos($x°$) = 0.8. What is the value of sin($90° − x°$)?

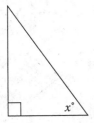

3. In the diagram below, Δ*JKL* is similar to Δ*QPR*, Δ*JKL*~Δ*QPR*. Which of the following is equal to tan(∠QPR)?

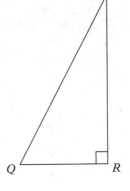

A. $\dfrac{JL}{JK}$

B. $\dfrac{JK}{JL}$

C. $\dfrac{JL}{KL}$

D. $\dfrac{KL}{JL}$

4. In a right triangle *ABC*, not shown, the hypotenuse is 10 and side *AB* = 6. What is the tangent of ∠C? Express your answer as a fraction.

5. In the diagram shown below, m∠ABC = 90°. Which of the following are equal to sin(∠BAD)?

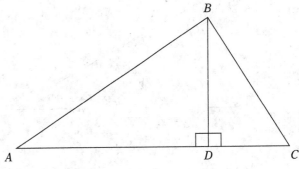

   I. cos(∠ABD)

  II. cos(∠DBC)

 III. cos(∠BCD)

A. I only

B. II only

C. I and III only

D. II and III only

6. In a right triangle, one angle measures $x°$, where sin $x° = \dfrac{12}{13}$. What is cos($90° − x°$)?

A. $\dfrac{12}{13}$

B. $\dfrac{5}{12}$

C. $\dfrac{12}{5}$

D. $\dfrac{13}{5}$

7. In right triangle *ABC* (not shown), angle *B* measures 60 degrees and the hypotenuse *BC* measures 12. What is tan(C) to the nearest hundredth?

8. Given that sin(x^2) = cos($x + 60$) and $x > 0$, what is a possible integer value of x?

9. In the diagram below, $\triangle ABC$ is similar to $\triangle DEF$, $\triangle ABC \sim \triangle DEF$. Find $\cos(\measuredangle F)$.

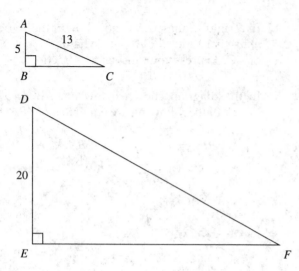

A. $\dfrac{5}{12}$

B. $\dfrac{12}{5}$

C. $\dfrac{12}{13}$

D. $\dfrac{5}{13}$

10. In the diagram below, $\triangle JKL$ is similar to $\triangle PQR$, $\triangle JKL \sim \triangle PQR$, $\tan(\measuredangle P) = \dfrac{4}{3}$, $PQ = 18$, and $KL = 8$. What is the length of $\overline{JL}$?

A. 6
B. 10
C. 24
D. 30

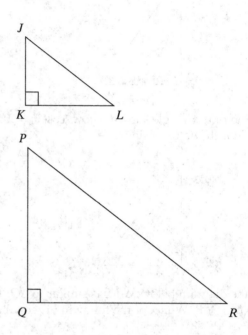

11. In the diagram below, $\sin\angle AED = \dfrac{12}{13}$, $BC = 10$, and $DE = 30$. What is the length of $\overline{CE}$?

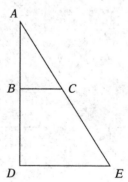

EXPLAINED ANSWERS

1. **EXPLANATION: The correct answer is 0.3.**

   Most problems of this type use SOHCAHTOA: cosine $= \dfrac{\text{adjacent}}{\text{hypotenuse}}$, but the SAT also wants you to know that the sine of one angle equals the cosine of another within a right triangle. If it helps, put in numbers for the sides: for example, side BC could be 3 and side AC could be 10 to get a sine of 0.3 as specified. You then have the sides you need to find the answer.

2. **EXPLANATION: The correct answer is 0.8.**

   Start by putting the sides in place for cosine of x: the adjacent (bottom) side is 8 and the hypotenuse (long) side is 10. Now determine which angle is $(90 - x)$. It's the angle in the other corner besides x, because the other two angles are already accounted for. Now use the opposite and hypotenuse sides of that angle to find that the sine of one angle is the same as the cosine of the given angle in the problem.

3. **EXPLANATION: Choice C is correct.**

   If the two triangles are similar, they necessarily have the same angles, so despite the fact that the bigger triangle has different sides, the tangent of an angle in that triangle will be the same as the tangent of the corresponding angle in the smaller triangle. In this case, that's angle J.

4. **EXPLANATION: The correct answer is $\dfrac{6}{8}$ or $\dfrac{3}{4}$.**

   Draw a picture. This is a 3-4-5 triangle, so if $AB = 6$, $BC = 8$. You can now find the tangent of $\angle C$, and don't forget that you can reduce the fraction if you want to when gridding it in, but it is not required. You'll still get the question right!

5. **EXPLANATION: Choice C is correct.**

   If this is hard to visualize, one way to do it is to set values for the sides of the triangles—just be sure you adhere to Pythagorean Theorem to be sure you're making real right triangles. For example, make side BD 4, DC 3, and BC 5. Side AD can be 8, and AB can be 10 (these are easy numbers to use, but any will work as long as they are an accurate right triangle). $\sin(\angle BAD)$ would then be $\dfrac{8}{10}$. Now check each of the Roman numerals.

   I. $\cos(\angle ABD) = \dfrac{8}{10}$. Eliminate B and D.

   II. $\cos(\angle DBC) = \dfrac{8}{4}$

   III. $\cos(\angle BCD) = \dfrac{8}{10}$

6. **EXPLANATION: Choice A is correct.**

   Draw the triangle: it's a 5-12-13, a Pythagorean triple. The three angles of the triangle are x, a 90-degree angle, and the third angle: this must be the $(90 - x)$ angle. The cosine of that angle will be 12 (adjacent) over 13 (hypotenuse).

7. **EXPLANATION: The correct answer is 0.58.**

   The problem tells you that this is a 30-60-90 right triangle. If the triangle is a 30-60-90, the small side is half of the hypotenuse, or in this case 6. The third side can be represented by s times the square root of 3. Taking the tangent of C means that we need to use the 6 side as the opposite side, and the other as the adjacent side: dividing the two numbers gives the credited response.

8. **EXPLANATION: The correct answer is 5.**

Remember that the sine of an angle equals the cosine of its complement. Therefore, $x^2 + (x + 60) = 90$. Solve for x by combining like terms and factoring: $(x + 6)(x - 5) = 0$. We need the positive value for x, so $x = 5$. Alternatively, use your calculator: $\cos(65) = 0.42$. $0.42 = \sin(x^2)$. If you're not sure where to go from here, don't forget that the answer is an integer. You can start plugging in—because we'll have to square it, this integer is likely to be fairly small. Use your calculator to help.

9. **EXPLANATION: Choice C is correct.**

The triangles are similar and the ratio of the small triangle's sides to the big triangle's sides is $1 : 4$, since $AB = 5$ and $DE = 20$. (To get from one triangle to the other, multiply by a factor of 4.) $\frac{5}{20} = \frac{13}{x}$. Crossmultiply to get $x = 52$. Side EF is 48, which you can determine either with the Pythagorean Theorem or by remembering that 5-12-13 is a Pythagorean triple, which will save you time. Now find $\cos(\angle F)$: $\frac{48}{52}$, which reduces to $\frac{12}{13}$. Alternatively, note that angle F corresponds to angle C. $BC = 12$, so $\cos F = \cos C = \frac{12}{13}$.

10. **EXPLANATION: Choice B is correct.**

The two triangles are similar, so their sides will be proportional. $\angle J$ will thus have the same tangent as $\angle P$ and we can use that fraction to extrapolate the sides. Side KL is 8 and is part of the tangent fraction, so figure out side JK: $\frac{4}{3} = \frac{8}{JK}$. $JK = 6$. This is a 3-4-5 triangle, a Pythagorean triple, but if you don't remember that, you can use the Pythagorean Theorem to determine that side $JL = 10$.

11. **EXPLANATION: The correct answer is 52.**

Remember that parts of similar triangles are proportional. Notice that triangle ADE is a 5-12-13. DE is the side that would represent 5, and 30 divided by $5 = 6$, so each of the sides of this triangle is the Pythagorean triple of 5-12-13 multiplied by 6. Therefore, $AE = 13 \times 6$, or 78. Now set up a proportion with the hypotenuses of the two triangles: $\frac{AC}{78} = \frac{10}{30}$. Crossmultiply to find that $AC = 26$. Now subtract that from AE: $78 - 26 = 52$, the length of CE.

CHAPTER 27

COMPLEX NUMBERS

On the SAT you will be required to simplify, rewrite, and identify the real and imaginary parts of a **complex number.** Begin with the mathematics review and then complete and correct the practice problems. There are 2 Solved SAT Problems and 11 Practice SAT Questions with answer explanations.

When trying to solve the equation $x^2 + 1 = 0$, first subtract 1 from both sides. This gives $x^2 = -1$. Squaring a number should always result in a positive value; therefore, there is no real number solution to $x^2 = -1$. We need to introduce a new type of number, $i = \sqrt{-1}$, which is the solution to the equation $x^2 = -1$.

A **complex number** is a number written in the form $a + bi$ where a and b are real numbers and $i = \sqrt{-1}$.

Since $i = \sqrt{-1}$. The important fact is that $i^2 = (\sqrt{-1})^2 \rightarrow i^2 = -1$

Example 1.

Simplify $(8i) \times (3i)$.

$(8i) \times (3i) = 24i^2 = 24(-1) = -24$

Example 2.

Simplify $(5 - 3i) + (-2 + 7i)$.

$(5 - 3i) + (-2 + 7i) = (5 - 2) + (-3 + 7)i = 3 + 4i$

Example 3.

Simplify $(6 + 4i) \times (2 - 5i)$.

$(6 + 4i) \times (2 - 5i) = (6 \times 2) + (6 \times -5i) + (4i \times 2) + (4i \times -5i)$

$= 12 - 30i + 8i - 20i^2$

$= 12 - 22i - 20(-1)$

$= 12 - 22i + 20$

$= 32 - 22i$

Example 4.

Simplify $\dfrac{8 + 3i}{6 - 2i}$.

First we need to multiply the numerator and denominator by the conjugate of the denominator. Since the denominator is $6 - 2i$, the conjugate is $6 + 2i$.

$$\dfrac{8 + 3i}{6 - 2i} \times \dfrac{6 + 2i}{6 + 2i} = \dfrac{48 + 16i + 18i + 6i^2}{36 + 12i - 12i - 4i^2}$$

$$= \dfrac{48 + 34i + 6(-1)}{36 + 0i - 4(-1)}$$

$$= \dfrac{48 + 34i - 6}{36 + 4} = \dfrac{42 + 34i}{40} = \dfrac{42}{40} + \dfrac{34}{40}i = \dfrac{21}{20} + \dfrac{17}{20}i$$

Practice Questions

1. Simplify $(-12 + 3i) + (7 + 6i)$.
2. Simplify $(8 - 4i) \times (3 + 5i)$.
3. Simplify $\dfrac{2 - 5i}{4 + 3i}$.

Practice Answers

1. $-5 + 9i$
2. $44 + 28i$
3. $\dfrac{-7}{25} - \dfrac{26}{25}i$

SOLVED SAT PROBLEMS

1. If the expression $(5 - 6i) \times (4 + 3i)$ is rewritten in the form $a + bi$ where a and b are real numbers, what is the value of b? (Note: $i = \sqrt{-1}$.)

   A. -3
   B. -6
   C. -9
   D. -27

   EXPLANATION: Choice C is correct.

   Follow the steps below:

   $$(5 - 6i) \times (4 + 3i) = 20 + 15i - 24i - 18i^2$$

   $$= 20 - 9i - 18(-1)$$

   $$= 20 - 9i + 18$$

   $$= 38 - 9i \quad \rightarrow \quad b = -9$$

2. Which of the following complex numbers is equivalent to $\frac{4+2i}{6-3i}$? (Note: $i = \sqrt{-1}$.)

A. $\frac{2}{3} - \frac{8}{9}i$

B. $\frac{2}{3} + \frac{8}{9}i$

C. $\frac{2}{5} - \frac{8}{15}i$

D. $\frac{2}{5} + \frac{8}{15}i$

EXPLANATION: Choice D is correct.

Follow the steps below:

$$\frac{4+2i}{6-3i} \times \frac{6+3i}{6+3i} = \frac{24 + 12i + 12i + 6i^2}{36 + 18i - 18i - 9i^2}$$

$$= \frac{24 + 24i + 6(-1)}{36 - 9(-1)}$$

$$= \frac{24 + 24i - 6}{36 + 9} = \frac{18 + 24i}{45} = \frac{18}{45} + \frac{24}{45}i = \frac{2}{5} + \frac{8}{15}i$$

COMPLEX NUMBERS
PRACTICE SAT QUESTIONS

ANSWER SHEET

Choose the correct answer.
If no choices are given, grid the answers in the section at the bottom of the page.

---

1. (A) (B) (C) (D)   11. GRID
2. GRID
3. (A) (B) (C) (D)
4. (A) (B) (C) (D)
5. (A) (B) (C) (D)
6. (A) (B) (C) (D)
7. GRID
8. (A) (B) (C) (D)
9. (A) (B) (C) (D)
10. (A) (B) (C) (D)

Use the answer spaces in the grids below if the question requires a grid-in response.

| Student-Produced Responses | ONLY ANSWERS ENTERED IN THE CIRCLES IN EACH GRID WILL BE SCORED. YOU WILL NOT RECEIVE CREDIT FOR ANYTHING WRITTEN IN THE BOXES ABOVE THE CIRCLES. |